T0134855

Springer Theses

Recognizing Outstanding Ph.D. Research

Aims and Scope

The series "Springer Theses" brings together a selection of the very best Ph.D. theses from around the world and across the physical sciences. Nominated and endorsed by two recognized specialists, each published volume has been selected for its scientific excellence and the high impact of its contents for the pertinent field of research. For greater accessibility to non-specialists, the published versions include an extended introduction, as well as a foreword by the student's supervisor explaining the special relevance of the work for the field. As a whole, the series will provide a valuable resource both for newcomers to the research fields described, and for other scientists seeking detailed background information on special questions. Finally, it provides an accredited documentation of the valuable contributions made by today's younger generation of scientists.

Theses are accepted into the series by invited nomination only and must fulfill all of the following criteria

- They must be written in good English.
- The topic should fall within the confines of Chemistry, Physics, Earth Sciences, Engineering and related interdisciplinary fields such as Materials, Nanoscience, Chemical Engineering, Complex Systems and Biophysics.
- The work reported in the thesis must represent a significant scientific advance.
- If the thesis includes previously published material, permission to reproduce this must be gained from the respective copyright holder.
- They must have been examined and passed during the 12 months prior to nomination.
- Each thesis should include a foreword by the supervisor outlining the significance of its content.
- The theses should have a clearly defined structure including an introduction accessible to scientists not expert in that particular field.

More information about this series at http://www.springer.com/series/8790

Jakub Sawicki

Delay Controlled Partial Synchronization in Complex Networks

Doctoral Thesis accepted by
Technische Universität Berlin, Germany

 Springer

Author
Dr. Jakub Sawicki
Institut für Theoretische Physik
Technische Universität Berlin
Berlin, Germany

Supervisor
Prof. Eckehard Schöll
Institut für Theoretische Physik
Technische Universität Berlin
Berlin, Germany

ISSN 2190-5053 ISSN 2190-5061 (electronic)
Springer Theses
ISBN 978-3-030-34078-0 ISBN 978-3-030-34076-6 (eBook)
https://doi.org/10.1007/978-3-030-34076-6

This Springer imprint is published by the registered company Springer Nature Switzerland AG
The registered company address is: Gewerbestrasse 11, 6330 Cham, Switzerland

Supervisor's Foreword

The focus of this thesis is on delay controlled partial synchronization patterns in complex networks. Synchronization is a ubiquitous phenomenon observed in different disciplines ranging from physics, acoustics, chemistry, biology, neuroscience, and engineering to socioeconomic systems. Synchronization of organ pipes is sometimes wanted and sometimes undesirable. Synchronization of power grids is necessary for a stable and robust operation, and failure may result in cascading power breakdown. Opinion formation in social networks or the stock prices in economic systems provide further examples. Fireflies synchronize their flashing light patterns as part of their mating display. Cardiac pacemaker cells fire synchronously in a regular heartbeat. The synchronization of neurons is believed to play a crucial role in the brain under normal conditions, for instance, in the context of cognition and learning, and under pathological conditions such as Parkinson's disease and epilepsy.

Of particular interest are the interplay of network topology, local nonlinear dynamics, and time-delayed couplings. Time delay is an efficient tool to control synchronization patterns. For example, the finite signal propagation time between coupled neurons significantly influences the dynamics. Besides complete synchronization (in-phase or anti-phase) or cluster and group synchronization, partial synchronization patterns like chimera states, which are characterized by the spatial coexistence of domains with coherent (synchronized) and incoherent (desynchronized) dynamics, are a key issue in current research. Moreover, many recent investigations have focussed upon multilayer systems, where nodes in different layers of the network are coupled in different ways. For instance, different remote layers which are not directly coupled may be synchronized via a relay layer, which itself obeys a different dynamics. This is called relay or remote synchronization.

This thesis aims for a fundamental understanding of these various synchronization phenomena, their interplay with the topology of complex networks, and their control via delayed coupling. It presents theoretical investigations and computer simulations, and is outstanding in its depth and breadth. Simple paradigmatic models are used for the local dynamics of the nodes: the Stuart-Landau model (normal form of a Hopf bifurcation), the Van der Pol oscillator (a generic nonlinear oscillator), the FitzHugh-Nagumo neuronal model (a paradigmatic model of neuronal excitability), and the time-discrete logistic map (a paradigm of chaos). After a short historical introduction and a concise survey of complex nonlinear systems and networks, the first part deals with single-layer system, and the second part presents multilayer systems.

In Part I of the thesis, at first the simplest configuration consisting of two delay-coupled nonlinear Van der Pol oscillators and its fascinating application to organ pipes, whose synchronization properties depend upon the distance, and hence upon the coupling delay, is considered analytically and numerically. The results offer an explanation of the surprising counter-intuitive nonmonotonic dependence of synchronization upon the distance of the organ pipes which have been found experimentally, but not been understood so far. Then chimera states in a ring network of nonlocally coupled Stuart-Landau oscillators are studied, explaining the chimera mechanism and the transition from classical phase chimeras to coupled amplitude-phase dynamics. Finally, the interplay of time delay and fractal network connectivities in ring networks of Van der Pol or FitzHugh-Nagumo oscillators is investigated, and typical resonance tongues of the chimera patterns are found. Fractal connectivities are motivated by empirical MRI data of the brain, revealing a fractal structure of the brain network.

Part II is concerned with multilayer structures. A neuronal model of the two brain hemispheres based upon empirical structural connectivities of the human brain is studied with the aim of explaining unihemispheric sleep. It has been speculated by many authors that chimera states may be related to unihemispheric sleep observed in migratory birds, aquatic mammals, and to a certain degree even in humans (*first night effect*), but this has never been proved by realistic models. Here for the first time, in computer simulations with empirical brain connectivities and FitzHugh-Nagumo dynamics, it is shown that in certain narrow parameter ranges of the interhemispheric coupling strength, chimera-like states of unihemispheric sleep can be obtained as a partial synchronization pattern. Finally, relay synchronization between remote layers in triplex networks is investigated. In brain research, many examples are known for relay functions of certain areas in the brain, for instance, the hippocampus serves as a relay between the frontal and the visual cortex. For the first time, partial relay synchronization patterns (double chimeras) are found in such a configuration with FitzHugh-Nagumo dynamics, i.e., only the coherent domains of the chimeras in the two remote layers are synchronized, but not the incoherent ones, and the relay layer exhibits completely different dynamics. Various partial synchronization scenarios are found in dependence of the interlayer coupling delay and the parameters of the two layers, and again a resonance tongue structure of the delay control is found. This offers an explanation of partial relay synchronization as observed in real brain experiments of mice. Such scenarios are universal, and they also occur in time-discrete coupled logistic maps. Universal mechanisms for partial relay synchronization can be related to intrinsic pacemakers.

This thesis gives a broad and comprehensive treatment of partial synchronization patterns and their control by time delay in complex networks. It studies and compares various network configurations and dynamic models, and breaks new grounds by focussing on novel partial synchronization scenarios and pointing out intriguing applications to musical acoustics (organ pipes) and neuroscience (uni-hemispheric sleep, relay effects in brain dynamics).

Berlin, Germany   Prof. Eckehard Schöll
September 2019

Abstract

The focus of this thesis are synchronization phenomena in networks and their intrinsic control through time delay. We encounter synchronization everywhere, and it can be a helpful as well as a detrimental mechanism, but definitely, a vital aspect for life on earth. As we are often not aware of it in small systems, it becomes fascinating in larger networks. Nevertheless, in both cases it is based on the same principles, which have been studied in some form for almost four centuries. For the emergence of synchronization, at least two oscillators, characterized by an intrinsic cycle, are necessary, together with a connection between them. We will focus on three generic models, namely the Stuart-Landau, Van der Pol, and FitzHugh-Nagumo system. Starting from a pair of oscillators and proceeding via simple ring networks, this thesis will outline the progression to complex multilayer structures.

Therefore, in Part I, a system of only two coupled oscillators is presented. By numerical and analytical analysis, we show that a seemingly simple system of two organ pipes gives birth to complex bifurcation and synchronization scenarios. Going from a two-oscillator system to a ring of oscillators, states between synchrony and desynchronization are thinkable, where only a part of the system synchronizes. An eminent example of partial synchrony are chimera states. For more than a decade, scientists have tried to solve the puzzle of this spontaneous symmetry breaking emerging on a ring of identical elements. Many previous works have covered various aspects of these states. We provide an analysis of initial conditions, and the understanding of the dependence on these could prove important. The model then is further extended by the addition of time delay, a crucial focus of this thesis. Delay is ubiquitous in the physical world and arises due to necessary propagation or processing time. The arising mathematical difficulties go hand in hand with the role of a powerful control mechanism.

In Part II, we proceed to multilayer systems. We investigate the occurrence of partial synchronization patterns in a neuronal network with empirical structural connectivity. Furthermore, we apply our findings to explain dynamical asymmetry arising from the hemispheric structure in human brain. A particular focus will be on the novel scenario of partial relay synchronization in multiplex networks, a special case of multilayer networks, where the individual layers represent different kinds of connections. As will be shown, triplex networks allow for synchronization of the coherent domains of chimera states in the outer layers via a remote layer, whereas the incoherent domains remain desynchronized. Finally, we develop an analytical approach and provide an explanation for the so-called double chimera states. Our results have implications for the understanding of partial synchronization in complex networks.

Acknowledgements

First and foremost, I would like to thank Prof. Eckehard Schöll in a fourfold way: As my advisor, he introduced me to the exciting field of network dynamics and supervised my doctoral research. Being the Coordinator of the SFB 910 during my time in the group, he gave me the opportunity to frequent visits of international conferences which enabled me to profit from the exchange of knowledge within the research community from a very early stage on. As a musician and poet, he inspired me and, last but not least, with his enthusiasm, he is an encouraging example and mentor.

I owe thanks to Dr. Anna Zakharova who provided guidance during the research and writing of this thesis. I am very much in debt to Dr. Judith Lehnert who was my first contact in the group and Dr. Iryna Omelchenko for her helpfulness. I am grateful to Rico Berner for insightful discussions. Furthermore, it would be remiss of me to neglect mentioning my students Fenja Drauschke, Stefan Apostel, Alexander zur Bonsen, Markus Galler, Johannes Meyer, David Mezey, Lukas Ramlow, Marius Winkler, and my RISE students Amy Searle, Natalia Spitha, and Nathan Myers. I owe thanks to all the current, former, and visiting members of the Schöll group for the fruitful working atmosphere. In particular, I thank for exciting discussions and collaborations in the framework of the SFB 910, namely Prof. Philipp Hövel, Prof. Serhiy Yanchuk, Dr. Simona Olmi, Dr. Thomas Isele, Dr. Benjamin Lingnau, Dr. Philipp Lorenz, Dr. Sergei Plotnikov, Dr. Andre Röhm, Dr. Jan Totz, Sanjukta Krishnagopal, Sarah Loos, Maria Masoliver, Liudmila Tumash, Roland Aust, Andreas Koher, Denis Nikitin, Leonhard Schülen, and Stefan Ulonska. Especially, I want to thank all musicians among my colleagues: Prof. Kathy Lüdge, Prof. Bernold Fielder, Dr. Jason Bassett, Mark Curran, and Nicola Vassena. I owe thanks to Yulia Jagodzinski, Andrea Schulze, and all secretaries for providing a fruitful environment in the framework of SFB 910 and TU Berlin.

I owe thanks to Prof. Markus Abel and Dr. Jost Fischer for introducing me to the topic of organ pipes and the cooperation, which gave me inspiration for Chap. 3. Furthermore, I would like to thank Peter Kalle for the fruitful collaboration which resulted in some of the content presented in Chap. 4. I would like to thank Prof. Wolfram Just. The collaboration with him and Dr. Paul Geffert was very pleasant and stimulating. I am also grateful for their great hospitality during my research visit to London. Likewise, I am grateful for the collaboration with Prof. Sarika Jalan and Saptarshi Ghosh. Their findings stimulated new numerical results which added to Chap. 7. I would also like to thank our Russian collaborators in the A1 project of the SFB 910, namely Prof. Vadim Anishchenko, Dr. Galina Strelkova, Dr. Nadezhda Semenova, Dr. Vladimir Semenov, and the entire group in Saratov.

I would like to thank Prof. Ralph Gregor Andrzejak for preparing the second assessment of this thesis and Prof. Dieter Breitschwerdt for chairing the defense of this thesis.

This work was funded by the Deutsche Forschungsgemeinschaft (DFG, German Research Foundation) – Projektnummer 163436311 – SFB 910 *Control of self-organizing nonlinear systems: Theoretical methods and concepts of application.* Furthermore, I acknowledge the support by the German Academic Exchange Service (DAAD) and the Department of Science and Technology of India (DST) within the PPP project (INT/FRG/DAAD/P-06/2018). This support enabled me to spend a research visit in the group of Prof. Sarika Jalan in Indore.

Contents

Chapter 1
Introduction

Relations form a fundamental category of human mind. In general, the awareness of relations requires the ability to compare and quantify. In the history of human kind this ability has an ambivalent aspect: On the one hand, it may be the basis for progress and prosperity, on the other hand, it may result in envy and ill-will. The extrinsic nature of relations is an interesting aspect, because the property "relation" itself exists only in dependence on its observer. As a musician and physicist I am often asked about my motivation for delving into these two rather different fields. But actually, there is indeed a deep relation between music and physics and in both fields the paradigm of relations had a significant impact on its evolution. The awareness for these relations can benefit from cross-pollination of ideas. It is not a coincidence that physicists like Huygens or Rayleigh, who have studied synchronization phenomena, have as well been dedicated musicologists. The aim of this introduction is to take a closer look on relations in and between music and physics. Our first approach is to compare two objects which are characterized by a change over time. These objects can be both musical and physical in nature, a musical instrument or a spring pendulum. In Sect. 1.1, we introduce the concept of synchronization, which means that the change over time of the two objects is correlated, or in other words, closely related to each other. The relation between the change itself and time will be characterized in Sect. 1.2. In case of more than two objects, we will take advantage of the network approach presented in Sect. 1.3. Section 1.4 gives the outline of the thesis.

1.1 Synchronization Phenomena

In 1665, Christiaan Huygens, sickly all his life, is ill in bed in his hometown Den Haag. A few years ago, he has invented the pendulum clock and its remarkable accuracy of within a few seconds a day has made him famous already. But beside his

© Springer Nature Switzerland AG 2019
J. Sawicki, *Delay Controlled Partial Synchronization in Complex Networks*,
Springer Theses, https://doi.org/10.1007/978-3-030-34076-6_1

Fig. 1.1 An original drawing of Christiaan Huygens illustrating his experiments in 1665 with two pendulum clocks mounted next to each other on the same brace [18]

illness, something else is causing a headache: Several sea trials have demonstrated to Huygens that the pendulum clock turned out to be much less accurate on the high seas. Even before this problem was reported, he has already had the idea to build a double-clock, just in case one of the clocks will stop working on a ship. Now, laying in his bed, he has mounted two of his pendulum clocks next to each other on the same brace (see Fig. 1.1) and accidentally observed "an odd kind of sympathy": Even if he disturbed one of the pendulums, after a short time, the clocks return to their original state swinging perfectly in opposite directions. This effect of entrainment was understood, centuries later, as *synchronization*.

Synchronization is a phenomenon that appears with a lot of facets in nature and fascinates human kind from centuries on. The Greek word *syn-chrónos* has the literal meaning of sharing the same time. That means, on the one hand, we need objects with well-defined intrinsic *time*, on the other hand, we need a shared connection between them. Synchronization occurs, when the intrinsic times of the single objects become *related*. This required constellation can be realized by oscillators which are coupled to each other. In case of a periodic oscillator the defined intrinsic timescale is called period. There are plenty of examples for synchronization patterns in nature and, especially in systems with a high number of elements they appear like magic, e.g., the flocking or swarm behavior of birds and fish, simultaneous flashing of thousands of fireflies, or the perfect interplay of musicians. Although these synchronization phenomena have been known for centuries, its systematical study started only in the twentieth century [37, 38]. The mechanism behind synchronization is a nonlinear one and the involved oscillators can be described by stable limit cycles in their so-called state space. These nonlinear oscillators are self-sustained and independent from an external force. From the energetic point of view, we are dealing with dissipative systems, where the energy is not conserved. We will focus on them in the next Sect. 1.2. Equally important for synchronization is the coupling between the oscillators. In the aforementioned example of Huygens clocks, the coupling is realized by the common brace. A system of coupled oscillators can be described by a network, where the oscillators are represented by nodes and the coupling between them by links (see Sect. 1.3). In general, to observe synchronization at least two coupled nonlinear oscillators are required. On the other side, there is no limit: Complex topologies and their interplay with complex local dynamics give rise to a plenty of synchronization scenarios. However, a universal feature of synchronization is the occurrence of a fixed relation between the individual time scales.

1.2 Dynamical Systems

The "revolution of the way of thinking" in Newtonian physics, as it was called by Immanuel Kant in his *Critique of Pure Reason*, is the cornerstone for the description of nonlinear oscillators. One could track this paradigm shift induced by Newton's description of nature in terms of *relations*: The basic idea of differential calculus is to provide the derivative of a function, which depends on a parameter. The derivative is given by the relation between the changes of the function, to the changes of the parameter. Knowing the relation of the changes gives us the opportunity to obtain a quantitative description of the function itself. Nowadays, dynamical systems in physics are described by such differential equations. Often it is impossible to find an explicit solution due to the high complexity of the systems of interest. In this case, differential equations serve as operating instructions for computer systems: Knowing the relation between the aforementioned changes and having a set of initial conditions, one is able to calculate step by step the solution numerically. This procedure requires a certain amount of computational power and this is why the history of nonlinear physics is closely related to the development of computers. The solution can be related to the state or phase space, where all possible states of the differential equation are represented, and the solution itself is described by a trajectory. A nonlinear, deterministic oscillator can be described by a closed trajectory, i.e. a self-repeating loop, in the phase space. A main aspect of nonlinear physics is to analyze these limit cycles under small perturbations. We can distinguish between stable and unstable dynamics. Moreover, if the solution of a differential system depends sensitively on initial conditions, we call its behavior chaotic.

In this thesis, we will consider generic models to study synchronization. As a model for oscillatory systems close to the Hopf bifurcation, we consider the Stuart-Landau oscillator [22]. As a model representing the nonlinear relaxation type oscillator, as opposed to sinusoidal oscillations, we will choose the Van der Pol model. The latter has been the first model where synchronization has been studied analytically [37, 38]. Furthermore, neural systems are described by excitable models. They can be classified according to their underlying bifurcation in type-I and type-II excitability (which will be described in detail later). As an example for type-II excitability we will investigate the FitzHugh-Nagumo system [14, 24]. Furthermore, we will investigate time-discrete dynamics. Thereby, we will consider the logistic map as a simple example which has been used in the literature to describe demographic evolution.

1.3 Network Description

In the previous Sections, we have introduced synchronization as a fixed relation between the temporal evolution of coupled oscillators. In case of two oscillators it is quite obvious to characterize a relation between them: Either there is one or there is none. Dealing with larger systems, however, it is more demanding to estimate relations between all oscillators. Moreover, we generally observe an additional amount of complexity when dealing with large systems composed of many related parts. "The whole is greater than the sum of its parts", as has been already realized by Aristotle. Nevertheless, large interconnected systems built from individual nodes with complex dynamics are common in many seemingly distinct fields of natural sciences, technology, and economy. A framework for this kind of problems is given by the field of network science, where connected oscillators are represented by links (connections) and nodes (oscillators) as shown in Fig. 2.6. In the human brain, for example, we are coming face to face with billions of nodes (neurons) connected to each other via as many as a quadrillion links. Interestingly, the first problem introduced in this field was actually a rather small one: The mathematician Leonhard Euler [13] asked himself if it would be possible to cross all bridges of Königsberg only once while returning to his starting point (Eulerian circuit). The considered network consists only of four nodes (land masses) and seven links (bridges). Nevertheless, the instruments provided by the graph theory – the mathematical equivalent of network science – simplifies this problem. It moreover provides solutions to much more complex questions. In fact, Euler stated that it is not possible to complete an Eulerian circuit if there are nodes of any odd degree, a network quantity characterizing the number of links of a node.

Today, networks have become an integral part of this world and their applications ranging from social science, economy, biology, chemistry, psychology, technology, mathematics, and physics [2, 6, 8, 26–28]. The growth of the world wide web is almost beyond comprehension, social networks register billions of daily active users, and search engines process over ten thousands of search queries every second, that means trillions of queries per year. Network scientist have developed various graph models to describe real-world systems, e.g., by random networks [29, 33], Erdős-Rényi networks [11, 12], small-world networks [1, 4, 5, 16, 17, 19, 23, 25, 32, 34–36, 39], scale-free networks [2, 3, 9, 10, 20], hierarchical networks [30], meta-networks, or interdependent networks. In particular, multilayer networks, networks with multiple kinds of relations, are quite topical [7, 15, 21, 31, 40]. Multilayer networks can give a general framework to describe and model real life examples, for instance the interactions between genes, proteins, neurons, transportation systems, power grids, and social networks. A common property of the mentioned examples is that they can be modeled with a network consisting of separate layers (multilayer network) in which the nodes are connected with different types of links (intra-layer) than those in between the layers (inter-layer). A special type of multilayer networks are multiplex networks, where each layer consists of the same set of nodes. The inter-layer connection is realized just between each node in one layer with its counterpart in the other layer. This network type is a focus of this thesis.

1.4 Outline of the Thesis

We now proceed by giving a brief outline. This thesis is divided into two main parts: Part I is centered on the study of synchronization phenomena in single-layer system, starting from a two-node system. Part II explores synchronization scenarios in multilayer networks as a common description of the neuronal brain structure. Parts of this thesis have been published (see list of Publications).

Before going into detail, we will discuss the general concepts and give an introduction to complex networks in Chap. 2. Part I begins with Chap. 3, where we give detailed account on a system of two coupled oscillators. We find various synchronization effects, develop analytical theory and apply it to acoustic experiments. As the focus of Chap. 4, we introduce networks with ring structures and explain special dynamical patterns called chimera states. We treat the impact of topology and initial conditions on the network dynamics. In Chap. 5, we investigate the situation where delay time enters into the system and reveal it as a powerful parameter to control dynamical patterns.

In Part II, we will proceed to multilayer systems. We start with Chap. 6 by considering empirical connectivity matrices which are closely related to the structure of the human brain. By analyzing partial synchronization patterns we give an explanation for an asymmetry in the special state of unihemispheric sleep. In Chap. 7, we consider an additional layer. A special focus is put on relay synchronization, where two outer layers are synchronized through a remote layer in between. For this, we explore double chimera states and realize control via time delay between and inside the three layers. These results are then generalized to different dynamical models and heterogeneities in the topology.

The thesis concludes with Chap. 8, where the obtained results are summarized and discussed. Furthermore, we give an outlook and consider future research directions.

References

1. Adamic LA (1999) The small world web, vol 1696/1999 of Lecture notes in computer science. Springer, Berlin
2. Albert R, Barabási AL (2002) Statistical mechanics of complex networks. Rev Mod Phys **74**:47–97
3. Barabási AL, Albert R (1999) Emergence of scaling in random networks. Science **286**:509
4. Barahona M, Pecora LM (2002) Synchronization in small-world systems. Phys Rev Lett **89**:054101
5. Bassett DS, Meyer-Lindenberg A, Achard S, Duke T, Bullmore ET (2006) Adaptive reconfiguration of fractal small-world human brain functional networks. Proc Natl Acad Sci USA **103**:19518–19523
6. Boccaletti S, Latora V, Moreno Y, Chavez M, Hwang DU (2006) Complex networks: Structure and dynamics. Phys Rep **424**:175–308
7. Boccaletti S, Bianconi G, Criado R, del Genio CI, Gómez-Gardeñes J, Romance M, Sendiña Nadal I, Wang Z, Zanin M (2014) The structure and dynamics of multilayer networks. Phys Rep **544**:1–122
8. Boccaletti S, Pisarchik AN, del Genio CI, Amann A (2018) Synchronization: from coupled systems to complex networks. Cambridge University Press, Cambridge
9. Chen Q, Chang H, Govindan R, Jamin S, Shenker SJ, Willinger W (2002) The origin of power laws in internet topologies revisited. In: Kermani P (ed) INFOCOM 2002. Twenty-first annual joint conference of the IEEE computer and communications societies. Proceedings, vol 2. IEEE, pp 608–617
10. de Solla Price DJ (1965) Networks of scientific papers. Science **149**:510–515
11. Erdős P, Rényi A (1959) On random graphs. Publ Math Debrecen **6**:290–297
12. Erdős P, Rényi A (1960) On the evolution of random graphs. Publ Math Inst Hung Acad Sci **5**:17–61
13. Euler L (1741) Solutio problematis ad geometriam situs pertinentis. Commentarii Acad Sci Petropolitanae **8**:128–140
14. FitzHugh R (1961) Impulses and physiological states in theoretical models of nerve membrane. Biophys J **1**:445–466
15. Gao J, Li D, Havlin S (2014) From a single network to a network of networks. Natl Sci Rev **1**:346–356
16. Hilgetag CC, Burns GAPC, O'Neill MA, Scannell JW, Young MP (2000) Anatomical connectivity defines the organization of clusters of cortical areas in the macaque and the cat. Philos Trans R Soc Lond Ser B **355**:91–110
17. Humphries MD, Gurney K, Prescott TJ (2006) The brainstem reticular formation is a small-world, not scale-free, network. Proc R Soc B Biol Sci **273**:503–511
18. Huygens C (1932) Oeuveres complétes de Christiaan Huygens, vol 17, includes works from 1651 to 1666. In: Nijhoff M (ed) Societe Hollandaise Des Sciences, La Haye
19. Jeong H, Tombor B, Albert R, Oltvai ZN, Barabási AL (2000) The large-scale organization of metabolic networks. Nature **407**:651
20. Jeong H, Mason SP, Barabási AL, Oltvai ZN (2001) Lethality and centrality in protein networks. Nature **411**:41–42
21. Kivelä M, Arenas A, Barthélemy M, Gleeson JP, Moreno Y, Porter MA (2014) Multilayer networks. J Complex Netw **2**:203–271
22. Kuramoto Y (1984) Chemical oscillations, waves and turbulence. Springer, Berlin
23. Monasson R (1999) Diffusion, localization and dispersion relations on "small-world" lattices. Eur Phys J B **12**:555–567
24. Nagumo J, Arimoto S, Yoshizawa S (1962) An active pulse transmission line simulating nerve axon. Proc IRE **50**:2061–2070
25. Newman MEJ, Watts DJ (1999) Renormalization group analysis of the small-world network model. Phys Lett A **263**:341–346

26. Newman MEJ (2003) The structure and function of complex networks. SIAM Rev **45**:167–256
27. Newman MEJ, Barabási AL, Watts DJ (2006) The structure and dynamics of networks. Princeton University Press, Princeton, USA
28. Newman MEJ (2010) Networks: an introduction. Oxford University Press Inc, New York
29. Rapoport A (1957) Contribution to the theory of random and biased nets. Bull Math Biol **19**:257–277
30. Ravasz E, Barabási AL (2003) Hierarchical organization in complex networks. Phys Rev E **67**:026112
31. Rheinwalt A, Goswami B, Boers N, Heitzig J, Marwan N, Kurths J (2014) A network of networks approach to investigate the influence of sea surface temperature variability on monsoon systems. In: EGU general assembly conference abstracts, vol 16. Springer, p 8147
32. Shefi O, Golding I, Segev R, Ben-Jacob E, Ayali A (2002) Morphological characterization of in vitro neuronal networks. Phys Rev E **66**:021905
33. Solomonoff R, Rapoport A (1951) Connectivety of random nets. Bull Math Biol **13**:107–117
34. Sporns O, Tononi G, Edelman GM (2000) Theoretical neuroanatomy: relating anatomical and functional connectivity in graphs and cortical connection matrices. Cereb Cortex **10**:127–141
35. Sporns O, Chialvo DR, Kaiser M, Hilgetag CC (2004) Organization, development and function of complex brain networks. Trends Cogn Sci **8**:418
36. Sporns O, Zwi JD (2004) The small world of the cerebral cortex. Neuroinformatics **2**:145–162
37. van der Pol B (1920) A theory of the amplitude of free and forced triode vibrations. Radio Rev **1**:701
38. van der Pol B (1926) On relaxation oscillations. Philos Mag **2**:978–992
39. Watts DJ, Strogatz SH (1998) Collective dynamics of 'small-world' networks. Nature **393**:440–442
40. Wiedermann M, Donges JF, Donner RV, Kurths J (2014) Ocean-atmosphere coupling from a climate network perspective. In: EGU general assembly conference abstracts, vol 16, p 11900

Chapter 2
Complex Systems

There is no clear definition of a complex system. One possible approach could be to characterize it by the ability of many interrelated elements to form a notable collective behavior. The reason for such an overall behavior cannot be explained either solely by the nonlinearity of the elements nor the diversity of the interrelations. It is rather an interplay of element and network properties, which gives rise to emergent behaviors like spontaneous order and self-organization. Therefore, complex systems can be seen as both, complex dynamics on networks, or dynamics on complex networks.

This Chapter is concerned with the general concepts of complex systems. In Sect. 2.1, we will discuss the common tools of nonlinear dynamics also known as chaos and bifurcation theory, whereas Sect. 2.2 gives an introduction to the mathematical description of networks.

2.1 Local Dynamics

Many systems in nature are inherently nonlinear. Since the beginning of the twentieth century, nonlinear oscillators have become an important research topic. Especially, in connection with the development of radio, radar (during World War II) and laser systems, the chaos theory has gained more and more attention. In the 1950s, the evolution of computer systems has established – next to theoretical and experimental physics – the field of computational or numerical physics. That progress has enabled the possibility to investigate new nonlinear models. In this Section, we will outline the main features of oscillator models in nonlinear dynamics. For a deeper understanding and more details we want to point out introductory literature, e.g., [36, 39, 46].

© Springer Nature Switzerland AG 2019
J. Sawicki, *Delay Controlled Partial Synchronization in Complex Networks*,
Springer Theses, https://doi.org/10.1007/978-3-030-34076-6_2

2.1.1 Differential Equations and Phase Space

Not only in physics, differential equations play a prominent role to describe dynamical systems. As mentioned in Sect. 1.2 they define the relation between quantities and their rate of change. Since Newton scientists have described many dynamical systems in nature by means of differential equations. In particular, deterministic time evolution of physical quantities $\mathbf{x}$ has been of great interest:

$$\dot{\mathbf{x}}(t) = \mathbf{f}\left(\mathbf{x}(t)\right). \tag{2.1}$$

Here, $\mathbf{x}$ denotes an n-dimensional vector of state variables and $\dot{\mathbf{x}}$ its derivative with respect to time t. The function $\mathbf{f}$ describes how the system state evolves in time. Since the last century the importance of differential equations even has grown in importance. The technological evolution has given birth to powerful computers, which can use even very complicated differential equation to calculate its solution step by step. This procedure is called numerical integration and the differential equation is in a sense the instruction for it. By using the definition of a differential quotient (calculus)

$$\dot{\mathbf{x}}(t) = \lim_{\Delta t \to 0} \frac{\mathbf{x}(t + \Delta t) - \mathbf{x}(t)}{\Delta t}, \tag{2.2}$$

and dividing the time in sufficiently small time steps Δt, one can map the present state $\mathbf{x}(t)$ to the next step $\mathbf{x}(t + \Delta t)$:

$$\mathbf{x}(t + \Delta t) = \mathbf{x}(t) + \mathbf{f}\left(\mathbf{x}(t)\right) \Delta t. \tag{2.3}$$

This iterative calculation is common in chaos theory, because it is not always possible to find an analytical solution for differential equations and give an explicit equation for $\mathbf{x}(t)$. Moreover, dynamical systems can also exhibit chaotic behavior: A small perturbation of the initial condition yields big differences of the solution. This is often the case in nonlinear systems, when the function $\mathbf{f}$ in Eq. (2.1) is dependent on $\mathbf{x}$ in a nonlinear way, e.g., $\mathbf{x}^k$ with higher power of $k > 1$. Each solution of a differential equation can be presented in the so-called phase space. It has been developed in the nineteenth century and represents every solution as a point in the space spanned by the n degrees of freedom. The time evolution of a solution is called phase space trajectory. Let us consider the Newtonian equation of motion for a simple, harmonic oscillator $\ddot{x} = -x$. This one-dimensional oscillator can be transformed to a dynamical system of the form of Eq. (2.1), where $\mathbf{x}$ is given by $\mathbf{x} = (x, \dot{x})^T$. In this case, the phase space with $n = 2$ is given by a phase plane $(x, \dot{x})$ as exemplarily shown in Fig. 2.1. Even if no analytical solution for a complicated Eq. (2.1) could be found, there is the possibility to determine trajectories of particular importance. Firstly, in case of

$$\dot{\mathbf{x}}^*(t) = 0 = \mathbf{f}\left(\mathbf{x}^*(t)\right). \tag{2.4}$$

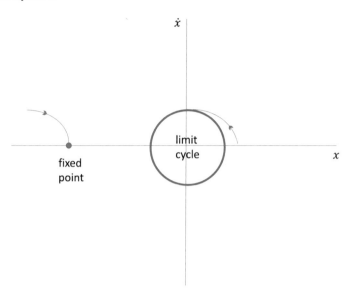

Fig. 2.1 Phase space as the figurative representation of dynamical systems: In the phase plane $(x, \dot{x})$ two invariant sets, a fixed point and a limit cycle, are shown. An exemplary trajectory is depicted in the neighborhood in case of a stable invariant set, respectively

we are speaking about fixed points $\mathbf{x}^*$ of the dynamical system. Starting the system from this point, the system remains there:

$$\dot{\mathbf{x}}^*(t) = 0 \quad \Rightarrow \quad \mathbf{x}^*(t) = \text{const.} \tag{2.5}$$

Secondly, there is the possibility of limit cycles. These trajectories are closed loops in the phase space and can be described by their period T_{LC}:

$$\mathbf{x}_{LC}(t) = \mathbf{x}_{LC}(t + T_{LC}). \tag{2.6}$$

Both invariant sets, a fixed point and a limit cycle are shown in Fig. 2.1 exemplarily. An important aspect in nonlinear dynamics is the stability of a trajectory Φ. For this purpose, arbitrary trajectories different from, but close to Φ are taken into account. If these trajectories remain close to Φ for all later times, Φ is called locally stable. If there exists a neighborhood of Φ, which consists of all points that approach Φ in the limit of $t \to \infty$, then Φ is called locally asymptotically stable and the corresponding neighborhood its basin of attraction. In the case of congruence between the whole phase space and the basin of attraction, Φ is denoted as globally stable. A bounded, undecomposable, invariant, and locally asymptotically stable subset of the phase space is called an *attractor* [39]. Both invariant sets mentioned above, the fixed point and the limit cycle, are attractors: the fixed point is a one-dimensional, the limit cycle a two-dimensional attractor. In the case of more than two dimensions, so-called chaotic attractors are possible, which are complicated sets in the phase space (e.g., the Rössler or Lorenz attractor). The opposite of an attractor is a repeller. In a more mathematical way one can quantify the stability with the instruments of linear stability analysis. For a small perturbation $\delta\mathbf{x}(0) = \mathbf{x}(0) - \mathbf{x}^*$ around a fixed point we are interested in its time evolution $\delta\mathbf{x}(t)$. Information about it can be achieved by using the first order term of the Taylor expansion for the time derivative of $\delta\mathbf{x}(t)$:

$$\dot{\delta\mathbf{x}}(t) = \delta\mathbf{x}(t)\mathbf{f}'(\mathbf{x}^*), \tag{2.7}$$

where $\mathbf{f}'$ is called the Jacobian matrix, an $n \times n$ matrix. With the exponential ansatz $\delta\mathbf{x}(t) = \delta\mathbf{x}(0)e^{\nu t}$ we obtain an eigenvalue problem of the Jacobian, where the spectrum of eigenvalues ν contains the essential information about the dynamical properties. The real parts of these in general complex eigenvalues are also known as Lyapunov-exponents $\Lambda = \mathrm{Re}(\nu)$ and we find that the largest Lyapunov-exponent Λ_{max} determines the stability of the system: the fixed point $\mathbf{x}^*$ is stable for $\Lambda_{max} < 0$ and becomes unstable for $\Lambda_{max} > 0$. The notion of Lyapunov exponents can be generalized to arbitrary trajectories and other (e.g., periodic or chaotic) attractors.

For limit cycles the Floquet theory answers the question about stability. Limit cycles are periodic attractors. In the following, we will give three examples of nonlinear oscillators used throughout this thesis. In these examples the function $\mathbf{f}$ in Eq. (2.1) is dependent on an additional bifurcation parameter $\mathfrak{p}$. For different values of this bifurcation parameter there exist different stability scenarios in the dynamical system. The point where the stability of a system changes drastically is called bifurcation point $\mathfrak{p}_0$. It is convenient to distinguish between different scenarios [39]:

Zero-eigenvalue bifurcation: This simplest local bifurcation occurs when a change of a system parameter $\mathfrak{p}$ causes a turn from positive to negative value of a single real eigenvalue. At the bifurcation point $\mathfrak{p}_0$ the eigenvalue is zero (center manifold). Examples for these bifurcations are the saddle-node, transcritical, super- and subcritical pitchfork bifurcation of fixed points.

Hopf bifurcation: This bifurcation occurs when a pair of complex-conjugate eigenvalues of a fixed point crosses the imaginary axis transversely. The fixed point changes from stable to unstable focus and a periodic attractor or limit cycle emerges at $\mathfrak{p}_0$ and its amplitude increases with $\sqrt{\mathfrak{p} - \mathfrak{p}_0}$. We can distinguish between the supercritical and subcritical Hopf bifurcation.

Local bifurcation of limit cycles: These bifurcations transform periodic attractors. Examples for such bifurcations are limit-cycle bifurcation by condensation of paths (stable and unstable limit cycle around the same fixed point coalesce), period-doubling bifurcation of a limit cycle (limit cycle with period twice the period of an original cycle replaces the latter one) and secondary Hopf bifurcation of a limit cycle (limit cycle is replaced by a two-dimensional torus).

Global bifurcation of limit cycles: Global bifurcations involve global changes of the topological configuration, e.g., when "larger" invariant sets, such as limit cycles, collide with fixed points. Examples for such bifurcations are the omega explosion, homoclinic, heteroclinic, or saddle-node infinite-period (SNIPER) bifurcation.

Bifurcation of spatial patterns: Considering high-dimensional dynamical systems, e.g., nonuniform spatial or spatio-temporal dissipative structures, there exists a wide variety of more complex bifurcations; among them scenarios like the Turing instability.

Bifurcations can be visualized by means of bifurcation diagrams. Such a diagram depicts fixed points or limit cycles as a function of a bifurcation parameter in the system as exemplarily shown in Fig. 2.5a.

Stuart-Landau oscillator

An important model for oscillators with coupled phase and amplitude dynamics is the Stuart-Landau system, which describes the behavior of a nonlinear oscillator near the Hopf bifurcation, i.e., the normal form of this bifurcation. Applications for this analytically tractable model can be found including in hydrodynamics, chemical reactions, and laser systems. The generalized equation of motion for the Stuart-Landau oscillator, named after Stuart [47] and Landau [25], reads

$$\dot{z} = \left(\lambda + i\omega - a_1 \left| z \right|^2 - a_2 \left| z \right|^4 \right) z, \tag{2.8}$$

where $z \in \mathbb{C}$ is a complex variable, $\lambda \in \mathbb{R}$ denotes the bifurcation parameter, and ω is the intrinsic frequency of the system. The real parameters a_1 and a_2 give the possibility to distinguish between the supercritical ($a_1 = 1, a_2 = 0$) and the subcritical ($a_1 = -1, a_2 = 1$) Hopf normal form. In case of $a_1 = a_2 = 0$ the system is linear and has no limit cycles, but a stable fixed point for $\lambda < 0$ (damped harmonic oscillations) and an unstable fixed point for $\lambda > 0$. By choosing polar coordinates for the complex variable $z = re^{i\phi}$ ($r \geq 0$), a decomposition into real and imaginary part is possible, which yields:

$$\begin{aligned} \dot{r} &= \lambda r + a_1 r^3 - a_2 r^5, \\ \dot{\phi} &= \omega. \end{aligned} \tag{2.9}$$

The invariant sets for $\lambda > 0$ in the supercritical case are given by

$$\begin{aligned} r_1^* &= 0, \\ r_2^* &= \sqrt{\lambda}, \end{aligned} \tag{2.10}$$

where r_1^* is an unstable fixed point and r_2^* a stable limit cycle. In the subcritical case the invariant sets are given by

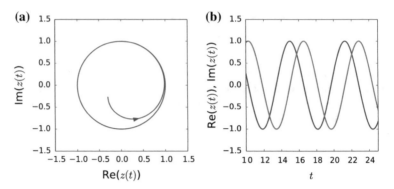

Fig. 2.2 Stuart-Landau system: **a** starting from an initial point, the trajectory of the variable $z(t) \in \mathbb{C}$ (blue line) of a Stuart-Landau oscillator in Eq. (2.8) approaches the limit cycle (in anti-clockwise direction); **b** time series for Re $(z(t))$ (dark blue line) and Im $(z(t))$ (light red line). Other parameters are given by $\lambda = 1.0$, $\omega = 1.0$, $a_1 = 1$ and $a_2 = 0$

$$r_1^* = 0,$$
$$r_{2/3}^* = \sqrt{\frac{1 \pm \sqrt{1 + 4\lambda}}{2}}, \tag{2.11}$$

where we have a stable $(+)$ and an unstable $(-)$ limit cycle. By a linear stability analysis one can distinguish between these stable and unstable solutions. A phase-portrait and time series of the complex state variable are shown in Fig. 2.2.

Van der Pol oscillator

Studying triode oscillations in electrical circuits, Balthazar van der Pol introduced the Van der Pol oscillator in 1927. Since that time his relaxation oscillator is one of the most intensely studied systems in nonlinear dynamics, e.g., seismology:

$$\ddot{x} - \varepsilon(1 - x^2)\dot{x} + x = 0. \tag{2.12}$$

In contrast to a linear oscillator, van der Pol replaced the damping coefficient by a nonlinear damping term $-\varepsilon(1 - x^2)$, where ε is a scalar parameter indicating the oscillators degree of non-linearity. The second-order differential equation can be rewritten in a two-dimensional form by the transformation $y = \dot{x}$:

$$\dot{x} = y,$$
$$\dot{y} = \varepsilon(1 - x^2)y - x. \tag{2.13}$$

In dependence on the nonlinear damping parameter ε one is able to tune the system from nearly sinusoidal to relaxation type oscillation with pronounced slow-fast dynamics. The Van der Pol oscillator has a stable fixed point at $x^*, y^* = 0$ for $\varepsilon < 0$

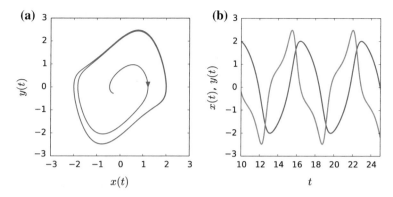

Fig. 2.3 Van der Pol system: **a** starting from an initial point, the trajectory in the phase plane of the variables $x(t)$ and $y(t)$ of a Van der Pol oscillator in Eq. (2.13) approaches the limit cycle (in clockwise direction); **b** time series for $x(t)$ (dark blue line) and $y(t)$ (light red line). The bifurcation parameter is given by $\varepsilon = 0.8$

and undergoes a Hopf bifurcation at $\varepsilon = 0$. A phase-portrait and time series of the state variables can be seen in Fig. 2.3.

FitzHugh-Nagumo oscillator

In the case of excitable systems we can distinguish between type-I and type-II excitability. The first one is characterized by the neighborhood of a saddle-node infinite period (SNIPER) bifurcation [39], also called the SNIC bifurcation (saddle-node bifurcation on invariant cycle). The second one arises close to a Hopf bifurcation. A common representative for this type is the FitzHugh-Nagumo system [14, 29]. Next to the Hodgkin-Huxley model [20, 21] it is a paradigmatic model to describe the dynamics of neural systems, but is also used to describe chemical [43] and optoelectronic [38] oscillators and nonlinear electronic circuits [19]. On the one hand, the FitzHugh-Nagumo oscillator can be seen as a simplification of the Hodgkin-Huxley model, since the ion channel gating variables are summarized in a single recovery variable (inhibitor), while the membrane potential corresponds to the activator. On the other hand, it can be derived by the aforementioned Van der Pol model. The nature of excitability is the possibility of the system to distinguish between small and large perturbations: For small perturbations we can observe small sub-threshold oscillations around the stable fixed point, whereas for larger perturbations prolonged excursions through the phase-space are triggered, the so-called super-threshold oscillations or spikes. This dual response-mechanism serves to describe the spiking behavior of neurons. In this thesis we will concentrate on the FitzHugh-Nagumo oscillator as a paradigmatic model for neuronal behavior, named after Richard FitzHugh and Jin-Ichi Nagumo. In contrast to the Hodgkin-Huxley model it has only two variables and therefore, allows for deeper analytical study:

$$\epsilon \dot{u} = u - \frac{u^3}{3} - v,$$

$$\dot{v} = u + a. \tag{2.14}$$

The state variable u is called activator and can be interpreted as the excitable membrane potential of neurons. By multiplying with the time-scale parameter $\epsilon \ll 1$, u becomes the fast variable compared to the other state variable v, which is called the inhibitor. Though there is no direct analogy of this recovery variable in terms of neuronal processes, the FitzHugh-Nagumo oscillator qualitatively reproduces the spiking behavior of the Hodgkin-Huxley model in general. Depending on the threshold parameter a, the FitzHugh-Nagumo oscillator exhibits excitable or oscillatory behavior. Linearizing around the fixed point $(u^*, v^*) = (-a, -a + a^3/3)$ yields a supercritical Hopf bifurcation for $|a| = 1$ as we can prove by calculating the eigenvalues λ:

$$\det \begin{pmatrix} 1 - a^2 - \lambda & 1 \\ -1 & -\lambda \end{pmatrix} = -\lambda(1 - a^2 - \lambda) + 1 = 0, \tag{2.15}$$

$$\Rightarrow \lambda_{1/2} = -\left(\frac{a^2 - 1}{2}\right) \pm \sqrt{\left(\frac{a^2 - 1}{2}\right)^2 - 1}. \tag{2.16}$$

For $|a| > 1$ the system has a stable fixed point and, therefore, remains in the excitable regime. By way of contrast, for $|a| < 1$ the unstable fixed point allows for self-sustained limit cycle oscillations. The transition between both regimes is not only governed by a Hopf bifurcation but a canard explosion which explains the very fast transition from a small amplitude limit cycle to a large amplitude relaxation oscillation [4]. A phase-portrait and time series of the state variables are shown in Fig. 2.4.

2.1.2 Time-Discrete Dynamical Systems

So far, we have introduced continuous-time systems. In many real-world problems we are faced with inherently discrete-time processes or the necessity to sample data. Especially, in nonlinear dynamics discrete-time processes play a prominent role. Named after Henri Poincaré, the transversal intersection of a periodic orbit of a continuous-time system with the Poincaré Section, a lower-dimensional subspace, is called Poincaré map. It maps the first intersection to the subsequent one and so on, and is also known as first recurrence map. Michel Hénon was one of the first investigating the motion of celestial objects by means of maps. Today the discretization of continuous-time systems is used to simplify the geometrical structure of chaotic attractors. Among other things, these recurrence relations can be applied to the evaluation of long-term electrocardiography.

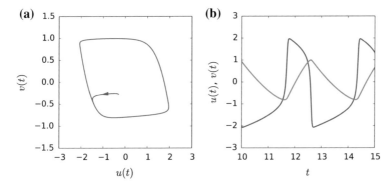

Fig. 2.4 FitzHugh-Nagumo system: **a** starting from an initial point, the trajectory in the phase plane of the variables $u(t)$ and $v(t)$ of a FitzHugh-Nagumo oscillator in Eq. (2.14) approaches the limit cycle (in anti-clockwise direction); **b** time series for $u(t)$ (dark blue line) and $v(t)$ (light red line). Other parameters are given by $\varepsilon = 0.05$ and $a = 0.5$

Logistic maps

The logistic map introduced by Pierre-François Verhulst in 1837 has been used inter alia to describe demographic evolution. The variable z represents the ratio of existing population to the maximum possible population and is recursively defined by

$$z_{n+1} = f(z_n) = az_n(1 - z_n). \tag{2.17}$$

This rather simple recurrence relation depends on the bifurcation parameter a, which has a great influence on the dynamics of the system. For $a < 1$ all orbits reach $z^* = 0$ and remain there. Between $1 \leq a \leq 3$ the system reaches one stable fixed point $z^* > 0$. As shown in Fig. 2.5a, for values $a > 3$ we observe that the single fixed point splits into two separate fixed points and after a period doubling bifurcation results in chaotic behavior. In the region $3.4 \leq a \leq 4$ where the single fixed point reaches a bifurcation point the attractor of the logistic map goes over to a period-2-cycle, as indicated by the two branches. To calculate the exact point of each period-doubling bifurcation the Lyapunov exponent can be used since it becomes zero at bifurcations (see for example the red dot in Fig. 2.5b for $a = 3$). The period-doubling cascade converges to $a \approx 3.57$ where the Lyapunov exponent Λ in Fig. 2.5b takes a positive value and the logistic map becomes chaotic [46]. In this region the logistic map is highly sensitive to initial conditions. Nevertheless, the grey area in Fig. 2.5a is interrupted by non-chaotic, periodic windows where several higher-period cycles exist for the logistic map dynamics.

Fig. 2.5 Bifurcation diagram of a logistic map: The upper panel **a** shows the behavior of the logistic map for values of a between $2.5 \leq a \leq 4$. We use 1000 iterations and plot the last 100 in the bifurcation diagram. Together with the Lyapunov exponent Λ in the lower panel **b** we can observe regions where the logistic map behaves chaotically (red) in dependence of a ($\Lambda > 0$)

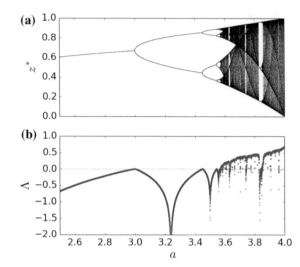

2.1.3   Delay Differential Equations

In Sect. 2.1.1, we have introduced differential equation as a common tool to model real-world processes. We can find them in physics, engineering, biology, medicine, economics, etc. and in the course of solving them one has to deal with an initial value problem. Equation (2.1) is a general example for such a so-called ordinary differential equation (ODE). In general, the change of a dynamical system does not depend solely on its current state, but often as well includes effects from its past. For instance, fishermen in South America have long observed the effects of El-Niño Southern Oscillation (ENSO). One reason of such a retarded effect might be propagation time [23]. In a system of two coupled oscillators the information needs a finite time to reach the second oscillator. So to speak, the second oscillator is coupled to the past of the first oscillator and vice versa. We can describe this situation by means of a delay differential equation (DDE):

$$\dot{\mathbf{x}}_i(t) = \mathbf{f}\left(\mathbf{x}_i(t)\right) + \sigma \mathbf{x}_j(t - \tau) \quad i, j = 1, 2, \tag{2.18}$$

where σ is the coupling strength and $\tau \geq 0$ is the delay time between node i and j. In general, there are no analytical instruments to solve delay differential equations, except in linear cases for which a transcendental characteristic equation can be derived. However, the main reason for the difficulties is the necessity of an initial history function to specify the value of the solution on the interval $[-\tau, 0]$. In contrast to the initial value problem of ordinary differential equations, in case of delay differential equations the derivative at any time depends on the solution at infinitely many prior times. Commonly, delay differential equations have to be treated numerically, where one has to define a specific history array (methods of steps). If the array is not chosen properly, we face the problem of jump derivative discontinuity at the initial time, which can propagate to future times and arises due to the incompatibility of the delay differential equation and the chosen initial history. During the propagation the discontinuity is smoothed out to higher order derivatives [31]. The history of the solution methods to these kinds of equations is also tightly connected to the evolution of computers. Therefore, the first numerical approach can be found in the seventies [30]. Today, there exists a wide range of literature on this topic [2, 11, 13, 17] and several recent Special Theme Issues have been published [16, 22, 48].

Time delay τ introduces an additional time-scale in the system, which can yield complex behavior due to the interplay with the intrinsic time-scale of the system. Moreover, time delay in a simple differential equation makes the equation infinite-dimensional in the sense that infinitely many initial conditions on the continuous interval $[-\tau, 0]$ are required. The increase of dimensionality often leads to chaotic behavior, as for example in the following equation [45]

$$\dot{y}(t) = \sin y(t - \tau), \tag{2.19}$$

where for $\tau > 0$ various bifurcation scenarios induced by time delay can be observed numerically. In contrast, for $\tau = 0$, the stability of the fixed points $y^* = m\pi$ ($m \in \mathbb{Z}$) of Eq. (2.19) can be calculated analytically. Proceeding to more complex systems, the interplay of time delay and topology and its influence on dynamics, in particular synchronization, is a recent topic of research [15, 18, 28, 40–42] and a main focus of this thesis.

2.2 Network Theory

In the previous Section, we have put the focus on the dynamics of oscillators. Although we have determined complex bifurcation behavior, we have basically dealt with single elements. As opposed to this, complex systems imply relations among these elements. In this Section, we will therefore focus on the theoretical foundations of network theory.

2.2.1 Network Topology and Adjacency Matrix

The basic idea of networks is to represent the relations between discrete objects. In the classical graph theory the objects are the vertices and the relation between them the edges. We will refer to the more physical terminology of nodes and links, respectively. For a given number N of nodes the distribution of links is determined by the network topology. A possible representation of the network topology is the $N \times N$ adjacency matrix $\mathbf{A}$:

$$\mathbf{A} = \{A_{ij}\} = \begin{pmatrix} A_{11} & \ldots & A_{1N} \\ \vdots & \ddots & \vdots \\ A_{N1} & \ldots & A_{NN} \end{pmatrix}, \tag{2.20}$$

where the individual elements are given by

$$A_{ij} = \begin{cases} 1, & \text{existing link from node } j \text{ to node } i, \\ 0, & \text{no link from node } j \text{ to node } i. \end{cases} \tag{2.21}$$

In general, the network topology can be directed or unidirectional, meaning, that the existence of a link from node i to node j does not imply a link from node j to node i, necessarily. Otherwise, the network is called undirected or bidirectional and therefore, the adjacency matrix is symmetric. In contrast to the binary adjacency matrix in Eq. (2.21), real-world networks need a gradation of the strength of a link. In that case one can distinguish between the unweighted binary matrix and a weighted matrix, where the value of A_{ij} indicates the strength of the link. The latter one is often referred to as the coupling matrix. Apart from Chap. 6, we will mainly focus on the unweighted type of adjacency matrices in this thesis.

By means of the adjacency matrix the dynamics on a network can be simply represented by an extension of Eq. (2.1) to a network term:

$$\dot{\mathbf{x}}_i = \underbrace{\mathbf{f}(\mathbf{x}_i)}_{\text{local term}} + \underbrace{\sigma \sum_{j=1}^{N} A_{ij}\mathbf{H}[\mathbf{x}_j - \mathbf{x}_i]}_{\text{network term}}, \tag{2.22}$$

where $\mathbf{x}_i \in \mathbb{R}^n$, $i = 1, \ldots, N$, is the local state of node i and $\mathbf{f}$ the local dynamics. The interaction is realized through diffusive coupling with coupling scheme $\mathbf{H}$ ($n \times n$ matrix) and coupling strength σ. In Fig. 2.6, the original bridge-problem of Euler is depicted as a graph. Actually, it is readily not possible to construct an adjacency

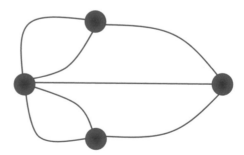

Fig. 2.6 First graph in history: In 1736, Leonhard Euler has solved a mathematical problem by means of graph theory and published a paper on *The Seven Bridges of Königsberg* providing a mathematical background on nodes (circles) and links (lines) [12]

matrix as in Eq. (2.20) for this graph. The reason is the multiple occurrence of the same link (multigraph). This issue can be solved in the framework of multilayer networks.

2.2.2 Multilayer Networks

In the previous Sect. 2.2.1, we have introduced a general description for networks by nodes and links. In this Section, we introduce a special kind of networks structures where it is practicable to define the notion "layer" in addition to nodes and links. Since the last decade scientists in various scientific fields have developed plenty of names to involve the layer-aspect into complex networks. Among them, multilayer network, multiplex network, multivariate network, multinetwork, multirelational network, multilayered network, multidimensional network, multislice network, hypernetwork, overlay network, composite network, multiweighted graph, multitype network, interconnected networks, interdependent networks, network of networks, and meta-network, to name just a few examples. In the most cases the terminology overlaps and is used inconclusively. In general, these networks are characterized by multiple levels or multiple types of links. As suggested in [24], we will use the multilayer network as a classificatory term denoting the extra notion of layer.

From the mathematical point of view it is not necessary to introduce an extra notion. Nevertheless, it may help to simplify complex structures and reveal basic mechanisms behind them. As an example, let us take a closer look on the spreading of diseases in a transportation network. The individual layers can be represented by an airline, a train, and a bus transportation layer. Because of the different speed of traveling on each layer, it is convenient to distinguish between those different layers and their time-scales. Another example are social networks: Here, the individual layers are represented by the different accounts, e.g., email, Twitter, Facebook, Instagram, Skype, etc. In this example, the transfer of information within a layer is in general faster than the transfer between the layers. Tracking a message becomes much easier by means of a multilayer description. Finally, we want to instance the last example: In an interconnected network like the connection between a power grid and a control network, one can often observe failure-cascades proceeding from one network to the other and back.

In Eq. (2.21), we have introduced the adjacency matrix for a single-layer network. In a similar way it is possible to define an adjacency tensor for multilayer networks:

$$A_{ij}^{\alpha\beta} = \begin{cases} 1, & \text{existing link from node } j \text{ in layer } \beta \text{ to node } i \text{ in layer } \alpha, \\ 0, & \text{no link from node } j \text{ in layer } \beta \text{ to node } i \text{ in layer } \alpha. \end{cases} \qquad (2.23)$$

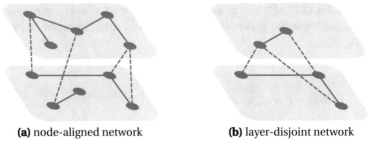

(a) node-aligned network   **(b)** layer-disjoint network

Fig. 2.7 Node-aligned (**a**) versus layer-disjoint (**b**) multilayer network: Two exemplary networks consist of two layers (light blue) with nodes (dark blue circles), connected via intra-layer (lines) and inter-layer links (dashed lines)

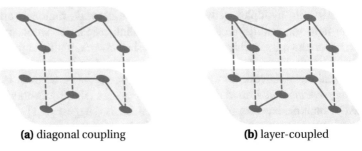

(a) diagonal coupling   **(b)** layer-coupled

Fig. 2.8 Diagonal (**a**) versus layer-coupled (**b**) multilayer network: Two exemplary networks consist of two layers (light blue) with each five nodes (dark blue circles), connected via intra-layer (lines) and inter-layer links (dashed lines)

To classify networks with multiple layers it is convenient to introduce following constraints:

- **node-aligned**: all of the layers contain all nodes as in Fig. 2.7a
- **layer-disjoint**: each node exists in at the most one layer as in Fig. 2.7b
- **diagonal coupling**: inter-layer links are only allowed between nodes and their counterparts in another layer as in Fig. 2.8a ($A_{ij}^{\alpha\beta} = 0$ if $i \neq j$ and $\alpha \neq \beta$)
- **layer-coupled**: links are independent of the nodes in diagonal coupled networks as in Fig. 2.8b ($A_{ii}^{\alpha\beta} = A_{jj}^{\alpha\beta}$)
- **categorical coupling**: each node is adjacent to all of its counterparts in the other layers as in Fig. 2.9a (subset of layer-coupled networks) ($A_{ii}^{\alpha\beta} = 1$ for $\alpha \neq \beta$)
- **ordinal coupling**: each node is adjacent to all of its counterparts in neighboring layers as in Fig. 2.9b

In case of a network which is not node-aligned, the tensorial description in Eq. (2.23) is faced with the problem of missing nodes in some layers. By adding empty nodes, which are not linked to any other nodes, one can resolve this trouble. A more elegant way is the representation by a supra-adjacency matrix. Such a matrix can be obtained by reduction of the rank of an adjacency tensor. Therefore,

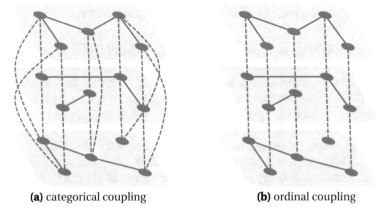

(a) categorical coupling   **(b)** ordinal coupling

Fig. 2.9 Categorical (**a**) versus ordinal coupled (**b**) multilayer network: Two exemplary networks consist of three layers (light blue) with each five nodes (dark blue circles), connected via intra-layer (lines) and inter-layer links (dashed lines)

two layers of the tensor are combined to one layer. This flattening process ultimately results in a supra-adjacency matrix $\mathbf{A}^{\mathscr{S}}$, where the representation of missing nodes is unproblematic. For a multilayer network with M layers the supra-adjacency matrix reads

$$
\mathbf{A}^{\mathscr{S}} = \begin{pmatrix} A_{ij}^{11} & A_{ij}^{12} & \cdots & A_{ij}^{1M} \\ A_{ij}^{21} & A_{ij}^{22} & \cdots & A_{ij}^{2M} \\ \vdots & \vdots & \ddots & \vdots \\ A_{ij}^{M1} & A_{ij}^{M2} & \cdots & A_{ij}^{MM} \end{pmatrix}, \tag{2.24}
$$

where $A_{ij}^{\alpha\beta}$ is given by Eq. (2.23).

Apart from the question of representation, one can roughly distinguish between node-colored and link-colored multilayer networks:

- **Node-colored networks** consist of two or more single-layer networks which are adjacent to each other via links that are called dependency links. The color of the nodes represents their label or type, e.g., sex or age in a social group. A further example is the dependency between a power grid and a computer network. In general, such networks are not node-aligned.
- **Link-colored networks** feature multiple types of links and are multilayer networks with diagonal coupling. An archetypical example for such networks are multiplex networks, which can represent for example social or transportation networks.

In Part II of this thesis, we will focus on node-aligned structures and, in particular, on multiplex networks.

2.2.3   Characteristic Quantities for Multiplex Networks

There is a broad range of approaches to quantify different aspects of complex networks [1, 5, 27, 32–34]. Most of them are based on a single-layer scenario and can be calculated from the adjacency matrix in Eq. (2.21). In this Section, which in parts follows [3, 6, 8], we introduce the most important measures for networks. Moreover, we demonstrate exemplarily, how these quantifications can be transferred or translated into the framework of multiplex networks.

Multiplex networks are sets of elementary units connected by links of different kinds. As a node-aligned network with a diagonal coupling structure (see previous Sect. 2.2.2), we can simplify the tensorial formulation of Eq. (2.23). In case of a multiplex network with categorical coupling the supra-adjacency matrix $\mathbf{A}_{cat}^{\mathscr{S}}$ is given by

$$
\mathbf{A}_{cat}^{\mathscr{S}} =
\begin{pmatrix}
A_{ij}^{11} & \mathbb{1} & \cdots & \mathbb{1} \\
\hline
\mathbb{1} & A_{ij}^{22} & \cdots & \mathbb{1} \\
\hline
\vdots & \vdots & \ddots & \vdots \\
\hline
\mathbb{1} & \mathbb{1} & \cdots & A_{ij}^{MM}
\end{pmatrix},
\tag{2.25}
$$

where $A_{ij}^{\alpha\alpha}$ is the coupling structure in layer α and $\mathbb{1}$ is the identity matrix. In case of a multiplex network with ordinal coupling the supra-adjacency matrix $\mathbf{A}_{ord}^{\mathscr{S}}$ simplifies further to

$$
\mathbf{A}_{ord}^{\mathscr{S}} =
\begin{pmatrix}
A_{ij}^{11} & \mathbb{1} & \cdots & \mathbb{0} \\
\hline
\mathbb{1} & A_{ij}^{22} & \cdots & \mathbb{0} \\
\hline
\vdots & \vdots & \ddots & \vdots \\
\hline
\mathbb{0} & \mathbb{0} & \cdots & A_{ij}^{MM}
\end{pmatrix},
\tag{2.26}
$$

where $\mathbb{0}$ is the zero matrix, which constitutes the most part of this supra-adjacency matrix. On the secondary diagonals we find the identity matrix $\mathbb{1}$.

Node degree: For single-layer networks, the node degree k_i is defined as the number of links connecting the node i:

$$
k_i = \sum_j A_{ij}.
\tag{2.27}
$$

In case of multilayer networks, this definition can be extended to the so-called *aggregated* network, where all layers α are quasi projected to a single layer:

$$k_i = \sum_j A_{ij}^{[\alpha]} \quad , \text{ with } A_{ij}^{[\alpha]} = \begin{cases} 1, & \text{if } \exists \alpha : A_{ij}^\alpha = 1, \\ 0, & \text{otherwise.} \end{cases} \quad (2.28)$$

In both cases, one can calculate the mean node degree by averaging over all nodes. Furthermore, for directed networks one can distinguish between in- and out-degree.

Overlapping degree: The overlapping degree o_i of node i is calculated by summing up the link overlaps o_{ij} for all links between node i and j:

$$o_i = \sum_j o_{ij} = \sum_j \sum_\alpha A_{ij}^\alpha = \sum_\alpha k_i^\alpha. \quad (2.29)$$

For a better comparison between multiplex networks of different size, it is recommendable to use the associated Z score:

$$Z(o_i) = \frac{o_i - \sum_i o_i}{\sigma_o}, \quad (2.30)$$

where σ_o is the corresponding standard deviation. For a Z score of $Z(o_i) \geq 2$, we can note the node as a hub, in contrast to a regular node for $Z(o_i) < 2$. A hub has a significantly larger number of links in comparison with a regular node.

Participation index: The node participation index P_i quantifies, whether the links of node i are uniformly distributed among the M layers of a multiplex network:

$$P_i = \frac{M}{M-1} \left[1 - \sum_\alpha \left(\frac{k_i^\alpha}{o_i} \right) \right], \quad (2.31)$$

where k_i^α is the node degree of node i in layer α. A lower value of this normalized measure means, that the node is rather connected within in layer than across the layers. A similar measure is the *layer and node pairwise multiplexity* [9, 35], which quantifies the correlation between the *activity* $\mathfrak{b}_i^\alpha$ of nodes in the layers. A node i without links in a layer α is called inactive ($\mathfrak{b}_i^\alpha = 0$), otherwise it is active ($\mathfrak{b}_i^\alpha = 1$).

Reachability: In a graph one can define the *shortest distance* (or geodesic distance) d_{ij} between node i and node j. To quantify the interdependency of a node in a multiplex network, it is interesting to know the shortest path ψ_{ij} between node i and node j, which make use of links in at least two layers. The reachability or interdependency of a node i is then given by

$$\lambda_i = \sum_{j \neq i} \frac{\psi_{ij}}{d_{ij}}. \tag{2.32}$$

For all of these measures the mean can be calculated by averaging over all nodes. The reachability is closely related to the *characteristic path length* [6] and the *efficiency* of a network [26], which can be also calculated by means of d_{ij} and generalized towards multilayer networks in a similar way.

Clustering coefficient: The global clustering coefficient C allows for characterizing the known phenomenon "the friend of your friend is my friend" by quantifying cliques in the network. We will introduce this coefficient for the case of one layer in all detail in Sect. 5.1. For a single node i the local clustering can be obtained by the ratio of the number of its triangles (triple of interconnected nodes i, j, k) and triads (correspond to triangles with no link between j and k):

$$C_i = \frac{\sum_{j \neq i, m \neq i} A_{ij} A_{jm} A_{mi}}{\sum_{j \neq i, m \neq i} A_{ij} A_{mi}}. \tag{2.33}$$

This definition can be modified for a multiplex network with M layers in various ways. We will focus on two possibilities, for which we need the following definition: A x-triangle/triad is a triangle/triad, whose links belong to x different layers of the network. The multiplex versions of the local clustering can be now given by

$$C_{i,x} = \frac{\text{number of } x+1\text{-triangles of node } i}{(M-x) \text{ number of } x\text{-triads of node } i}, \quad x = 1, 2. \tag{2.34}$$

The global clustering coefficient of the system is obtained by averaging over all nodes i.

Transitivity: The transitivity T is closely related to the clustering coefficient. Instead of taking into account every single node, the ratio of cliques in the whole network is calculated (see Sect. 5.1). For a multiplex network with M layers the transivity can be defined as follows:

$$T_x = \frac{\text{number of } x+1\text{-triangles in the graph}}{(M-x) \text{ number of } x\text{-triads in the graph}}, \quad x = 1, 2. \tag{2.35}$$

Eigenvector centrality: The eigenvector centrality of a node i can be obtained from the ith component of the eigenvector related to the leading eigenvalue of an adjacency matrix. A high eigenvector centrality means that the corresponding node is connected to many nodes who themselves have a high centrality. In a multiplex network with M layers, we have the possibility to investigate the influence of each layer on the eigenvector centrality of a node, separately. By a linear combination of the adjacency matrices of each layer $\mathbf{A}^\alpha$ it is possible to determine the contribution of each layer to the centrality of a node [3, 44]. This combination of adjacency matrices is called multiadjacency matrix $\mathbf{A}_{\mathcal{M}}$ and can be obtained as follows:

$$\mathbf{A}_{\mathcal{M}} = \sum_{\alpha=1}^{M} b_\alpha \mathbf{A}^\alpha \quad , \sum_{\alpha=1}^{M} b_\alpha = 1. \tag{2.36}$$

Many applications, such as ranking algorithms for websites, are interested in the dominant eigenvalue of the adjacency matrix. Nevertheless, also the other eigenvalues provide valuable information about structure and dynamics of the network [8]. The whole set of eigenvalues of an adjacency matrix is called its *spectrum* and can be obtained – in case of multiplex network – either from the aforementioned aggregated system (see node degree) or from the multiadjacency matrix $\mathbf{A}_{\mathcal{M}}$.

2.2.4 Synchronization and Master Stability Equation

In Sect. 1.1, we have introduced the phenomenon of synchronization as a fixed relation between dynamical objects. From the mathematical point there exist many possibilities to define synchronization. In this thesis, we will deal mainly with oscillatory systems, in which a phase ϕ can be defined. Thus, phase-lag synchronization is an appropriate basis: Two oscillators i and j are phase-lag synchronized, if

$$x_i(t) = x_j(t + \tfrac{\phi}{2\pi}T). \tag{2.37}$$

In oscillatory systems T represents the period of the oscillations. Then, in case of $\phi = 0$ we use the term *in-phase* synchronization, whereas for $\phi = \pi$ we use the term *anti-phase* synchronization. Dealing with chimera states we often rely on the mean phase velocity of an oscillator without tracking its exact timeseries. In such a case we will refer to the more general definition of *frequency* synchronization, for which the mean phase velocities of two oscillators coincide (see Sect. 4.1.1).

In Eq. (2.22), a network with diffusive coupling is given. We assume in-phase synchronization $\mathbf{x}_i = \mathbf{x}_j = \mathbf{x}_s$ and are interested in the stability of this synchronized state. In Sect. 2.1.1, we have introduced the stability analysis for dynamical systems, which can be applied also in this case: By the linearization of the network dynamics at the synchronous state

$$\dot{\mathbf{x}}_s = \mathbf{F}(\mathbf{x}_s) + \sigma \sum_{j=1}^{N} A_{ij} \mathbf{H}[\mathbf{x}_s - \mathbf{x}_s], \tag{2.38}$$

one can then calculate the largest Lyapunov exponent Λ_{max} to determine the stability. Depending on network size this approach can be a quite computational demanding procedure. A more elegant path embarks the master stability equation invented by Pecora and Carroll [37], which has been generalized to delay-coupled networks in [7]. Equation (2.22) can be rewritten in terms of matrices by defining $\mathbf{X} = \{\mathbf{x_i}\}$:

$$\dot{\mathbf{X}} = \mathbf{F}(\mathbf{X}) + \sigma \mathbf{A} \otimes \mathbf{H}(\mathbf{X}), \tag{2.39}$$

where $\mathbf{F}(\mathbf{X}) = \{\mathbf{f}(\mathbf{x}_i)\}$, $\mathbf{H}(\mathbf{X}) = \{\mathbf{H}(\mathbf{x}_i)\}$, and $\otimes$ stands for the tensor product. Equivalently to Eq. (2.7) the variational equation for a small perturbation around the synchronized state $\mathbf{X}_s + \delta \mathbf{X} = \{\mathbf{x}_s + \delta\mathbf{x}_i\}$ reads

$$\delta\dot{\mathbf{X}} = [DF(\mathbf{X}_s) + \sigma \mathbf{A} \otimes \mathbf{H}]\delta\mathbf{X}. \tag{2.40}$$

The crucial step is the diagonalization of the adjacency matrix $\mathrm{diag}(\mathbf{A}) = \mathbf{S}^{-1}\mathbf{A}\mathbf{S}$, where $\mathbf{S}^{-1}$ is the diagonalizing matrix of $\mathbf{A}$ with $\mathbf{S}^{-1}\mathbf{S} = \mathbb{1}$. By means of the transformation $\delta\tilde{\mathbf{X}} = (\mathbf{S}^{-1} \otimes \mathbb{1}_m)\delta\mathbf{X}$, where m is the dimension of the local phase space, we obtain the master stability equation

$$\delta\dot{\tilde{\mathbf{x}}}_i = [Df(\mathbf{x}_s) + \sigma \lambda_i \mathbf{H}]\delta\tilde{\mathbf{x}}_i, \tag{2.41}$$

where λ_i are the eigenvalues of the coupling matrix $\mathbf{A}$. Now, instead of Eq. (2.38) the master stability equation (2.41) can be taken to calculate the maximum Lyapunov exponent Λ_{max} and determine the stability of the synchronized state. Thus, the idea of the master stability function is the separation of the effects of local dynamics from the effects of the topology. In this derivation, we have used the assumption of identical local dynamics $\mathbf{f}$ and coupling scheme $\mathbf{H}$, which is not obligatory as has been shown in [10, 28] for networks with a delayed coupling. Moreover, the row-sum of the coupling matrix $\mathbf{A}$ has to be constant ($\sum_j A_{kj} = $ const.). This constraint holds in case of circulant matrices, where each row of the coupling matrix is rotated one element to the right relative to the preceding row.

2.3　Summary

The focus of this Chapter have been the theoretical foundations of complex systems. We have considered them from the perspective of nonlinear dynamics on complex networks. Section 2.1 has been concerned with the dynamical aspect. As a common description for the dynamics of nonlinear systems, we have introduced differential equations and, in particular delay differential equations, which meet the needs of real world problems. A prominent example of complex dynamics can be found in the bifurcation scenarios of nonlinear oscillators. Therefore, after giving a short explanation of stability and bifurcation theory in dynamical systems, we have introduced three paradigmatic models, which have a wide range of applicability in different scientific fields as electronic circuits, chemical reactions, and neuronal dynamics.

Section 2.2 has taken the network aspect under account. We have introduced the mathematical framework to describe networks and provide characteristic quantifications to have an overview of their properties. Moreover, we have presented possibilities to transfer these quantities to multilayer and, in particular, to multiplex networks. These types of networks can be classified by the concept of layers as an additional dimension next to node and links. We have provided a terminology to characterize and categorize different types of multilayer and multiplex networks. Finally, we have introduced the master stability equation to determine synchronization phenomena in arbitrary networks. Before proceeding to complex networks, we will first discuss synchronization scenarios in the smallest possible network of two coupled nonlinear oscillators in the following Chapter.

References

1. Albert R, Barabási AL (2002) Statistical mechanics of complex networks. Rev Mod Phys **74**:47–97
2. Atay FM (ed) (2010) Complex time-delay systems, understanding complex systems. Springer, Berlin
3. Battiston F, Nicosia V, Latora V (2014) Structural measures for multiplex networks. Phys Rev E **89**:032804
4. Benoit EE, Callot JL, Diener F, Diener MM (1981) Chasse au canard (première partie). Collect Math **32**:37–119
5. Boccaletti S, Latora V, Moreno Y, Chavez M, Hwang DU (2006) Complex networks: structure and dynamics. Phys Rep **424**:175–308
6. Boccaletti S, Bianconi G, Criado R, del Genio CI, Gómez-Gardeñes J, Romance M, Sendiña Nadal I, Wang Z, Zanin M (2014) The structure and dynamics of multilayer networks. Phys Rep **544**:1–122
7. Choe CU, Dahms T, Hövel P, Schöll E (2010) Controlling synchrony by delay coupling in networks: from in-phase to splay and cluster states. Phys Rev E **81**:025205(R)
8. Cozzo E, De Arruda GF, Rodrigues FA, Moreno Y (2018) Multiplex networks: basic formalism and structural properties. Springer, Berlin
9. Criado R, Flores J, GarcĀa del Amo A, Gómez-Gardeñes J, Romance M (2012) A mathematical model for networks with structures in the mesoscale. Int J Comput Math **89**:291
10. Dahms T, Lehnert J, Schöll E (2012) Cluster and group synchronization in delay-coupled networks. Phys Rev E **86**:016202
11. Erneux T (2009) Applied delay differential equations. Springer, Berlin
12. Euler L (1741) Solutio problematis ad geometriam situs pertinentis. Commentarii Acad Sci Petropolitanae **8**:128–140
13. Farmer JD (1982) Chaotic attractors of an infinite-dimensional dynamical system. Phys D **4**:366
14. FitzHugh R (1961) Impulses and physiological states in theoretical models of nerve membrane. Biophys J **1**:445–466
15. Flunkert V (2011) Delay-coupled complex systems, Springer theses. Springer, Heidelberg
16. Flunkert V, Fischer I, Schöll E (2013) Dynamics, control and information in delay-coupled systems. Theme Issue of Phil Trans R Soc A **371**:20120465
17. Fridman E (2014) Introduction to time-delay systems: analysis and control. Springer, Berlin
18. Hövel P (2010) Control of complex nonlinear systems with delay, Springer theses. Springer, Heidelberg

19. Heinrich M, Dahms T, Flunkert V, Teitsworth SW, Schöll E (2010) Symmetry breaking transitions in networks of nonlinear circuit elements. New J Phys **12**:113030
20. Hodgkin AL (1948) The local electric changes associated with repetitive action in a medullated axon. J Physiol **107**:165
21. Hodgkin AL, Huxley AF (1952) A quantitative description of membrane current and its application to conduction and excitation in nerve. J Physiol **117**:500–544
22. Just W, Pelster A, Schanz M, Schöll E (2010) Delayed complex systems. Theme Issue of Phil Trans R Soc A **368**:301–513
23. Keane A, Krauskopf B, Postlethwaite CM (2017) Climate models with delay differential equations. Chaos **27**:114309
24. Kivelä M, Arenas A, Barthélemy M, Gleeson JP, Moreno Y, Porter MA (2014) Multilayer networks. J Complex Netw **2**:203–271
25. Landau LD (1944) On the problem of turbulence. C R Acad Sci UESS **44**:311
26. Latora V, Marchiori M (2001) Efficient behavior of small-world networks. Phys Rev Lett **87**:198701
27. Lehnert J (2010) Dynamics of neural networks with delay. Master's thesis, Technische Universität Berlin
28. Lehnert J (2016) Controlling synchronization patterns in complex networks, Springer theses. Springer, Heidelberg
29. Nagumo J, Arimoto S, Yoshizawa S (1962) An active pulse transmission line simulating nerve axon. Proc IRE **50**:2061–2070
30. Neves KW (1975) Automatic integration of functional differential equations: an approach. ACM Trans Math Softw **1**:357
31. Neves KW, Feldstein A (1976) Characterization of jump discontinuities for state dependent delay differential equations. J Math Anal Appl **5**:689
32. Newman MEJ (2003) The structure and function of complex networks. SIAM Rev **45**:167–256
33. Newman MEJ, Barabási AL, Watts DJ (2006) The structure and dynamics of networks. Princeton University Press, Princeton, USA
34. Newman MEJ (2010) Networks: an introduction. Oxford University Press Inc, New York
35. Nicosia V, Latora V (2015) Measuring and modeling correlations in multiplex networks. Phys Rev E **92**:032805
36. Ott E (2002) Chaos in dynamical systems. Cambridge University Press, Cambridge
37. Pecora LM, Carroll TL (1998) Master stability functions for synchronized coupled systems. Phys Rev Lett **80**:2109–2112
38. Rosin DP, Callan KE, Gauthier DJ, Schöll E (2011) Pulse-train solutions and excitability in an optoelectronic oscillator. Europhys Lett **96**:34001
39. Schöll E (2001) Nonlinear spatio-temporal dynamics and chaos in semiconductors. Nonlinear science series, vol 10. Cambridge University Press, Cambridge
40. Schöll E, Schuster HG (eds) Handbook of chaos control. Second completely revised and enlarged edition. Wiley-VCH, Weinheim
41. Schöll E (2013) Synchronization in delay-coupled complex networks. In: Sun J-Q, Ding Q (Eds) Advances in analysis and control of time-delayed dynamical systems, Chapter 4. World Scientific, Singapore, pp 57–83
42. Schöll E, Klapp SHL, Hövel P (2016) Control of self-organizing nonlinear systems. Springer, Berlin
43. Shima S, Kuramoto Y (2004) Rotating spiral waves with phase-randomized core in nonlocally coupled oscillators. Phys Rev E **69**:036213
44. Solá L, Romance M, Criado R, Flores J, Garcia del Amo A, Boccaletti S (2013) Eigenvector centrality of nodes in multiplex networks. Chaos **23**:033131
45. Sprott JC (2007) A simple chaotic delay differential equation. Phys Lett A **366**:397
46. Strogatz SH (1994) Nonlinear dynamics and chaos. Westview Press, Cambridge, MA
47. Stuart JT (1958) On the non-linear mechanics of hydrodynamic stability. J Fluid Mech **4**:1
48. Sun JQ, Ding G (2013) Advances in analysis and control of time-delayed dynamical systems. World Scientific, Singapore

Part I
Single-Layer Systems

Chapter 3
Two Coupled Oscillators

In this Chapter, we investigate synchronization of two coupled oscillators using the example of organ pipes. It is well-known that synchronization and reflection in the organ lead to undesired weakening of the sound in special cases. Recent experiments have shown that sound interaction is highly complex and nonlinear. However, we show that already two delay-coupled Van der Pol oscillators in fact appear to be a good model for the occurring dynamical phenomena. Here, the coupling is realized as distance-dependent, or time-delayed, equivalently. We analytically investigate the synchronization frequency and bifurcation scenarios which occur at the boundaries of the Arnold tongues. We successfully compare our results to experimental data. The following Chapter is closely related to my publications [19–21].

This Chapter is structured as follows: First, we explain the nonlinear dynamical behavior of the organ in Sect. 3.1. In Sect. 3.2, we introduce two delay-coupled Van der Pol oscillators as a simple model of coupled organ pipes. In Sect. 3.3, we apply two analytical methods to get a better understanding of the synchronization phenomena. The central part of the Chapter is Sect. 3.4, where we present the analytical results. In Sect. 3.5, we compare these results with acoustic experiments, and in Sect. 3.6, we discuss the importance of an appropriate coupling function. We conclude with Sect. 3.7.

3.1 The Organ Pipe – a Nonlinear Oscillator

The physics of organ pipes is an interdisciplinary topic where many fields of science meet. It is highly interesting as it includes elements of nonlinear dynamical system theory [4, 6, 10], aeroacoustic modeling [13] and synchronization theory [17, 24]. The focus of these different research areas is the "queen of instruments" which captivates through the grandeur of her sight and majesty of her sound (see Fig. 3.1).

© Springer Nature Switzerland AG 2019
J. Sawicki, *Delay Controlled Partial Synchronization in Complex Networks*,
Springer Theses, https://doi.org/10.1007/978-3-030-34076-6_3

Fig. 3.1 The "queen of instruments": the impressive organ of the Berlin Cathedral build by Wilhelm Sauer in 1905 with in total 7269 pipes. The largest preserved organ in its original state – dating back to the 'Late Romantic' period – has a width of 14 m and height of about 20 m. Picture presented by kind permission of Maren Glockner

Here, we investigate an interesting nonlinear effect: organ pipes close to each other synchronize. Recent studies have been of experimental nature as well as theoretical [1, 2, 7, 8]. For musical purposes, synchronization of sound might be desired or not: it might stabilize the pitch of special organ pipes as a favorable effect, whereas sound weakening, as observed in the prospect of an organ, is highly undesired [9, 18, 23]. This weakening occurs as an amplitude minimum due to destructive interaction between pipes, e.g., during the actuating of the swell box, where the pipes stand close to each other.

A qualitative understanding of the nonlinear mechanisms is obtained following the arguments of [2, 7]: a single organ pipe can be described as a self-sustained oscillator, where the oscillating unit consists of the jet, or "air sheet", which exits at the pipe mouth. The resonator is, of course, given by the pipe body; there, sound waves emitted at the labium (i.e., the sharp edge in the upper part of the pipe's opening) travel up and down and can trigger a regular oscillation of the air sheet. Energy is supplied by the generating unit, which is the pressure reservoir beneath the pipe at a basically constant rate. The schematic structure of a closed flue pipe is

Fig. 3.2 Organ pipe as a
self-sustained oscillator:
The pressure waves driven
by the supplied airstream
trigger the "air sheet" which
exits at the pipe mouth.
Figure modified from [19]

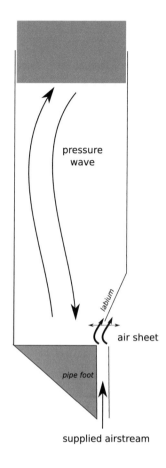

pressure
wave

labium

air sheet

pipe foot

supplied airstream

shown in Fig. 3.2. Experimental and numerical investigations by Abel et al. [2] yield the conclusion that an organ pipe can be approximated satisfactorily by a Van der Pol oscillator.

While the work of Fischer [7] focuses on the nonlinearities in sound generation and their effect on the synchronization properties, here we investigate the effect of the finite distance of two coupled pipes which, in turn, is reflected by a delay in the coupling function. More specifically, we analyze the bifurcation scenarios in the context of two delay-coupled Van der Pol oscillators as a representation of the system of two coupled organ pipes, such as in the experimental setup of Bergweiler et al. [5]. In extension of previous work, we study the dependence of Arnold tongues under variation of the time delay τ and the coupling strength κ, to explore how undesired synchronization or chaotic behavior can be avoided. We compare our results to experimental measurements of the synchronization under variation of the pipe distance. Indeed, we find a qualitative coincidence of the nonmonotonic modulation of the shape of the Arnold tongue which contrasts the linear boundaries of the Arnold tongues for systems with undelayed coupling.

3.2 A Model of Coupled Organ Pipes

To gain a deeper insight into the synchronization phenomena of two coupled organ
pipes, we model the pipes by Van der Pol oscillators with delayed cross-coupling:

$$\ddot{x}_i + \omega_i^2 x_i - \mu \left[\dot{x}_i - \dot{f}(x_i) + \kappa(\tau)x_j(t - \tau) \right] = 0, \tag{3.1}$$

where $i, j = 1, 2$. These equations represent a harmonic oscillator with an intrinsic
angular frequency ω_i, supplemented with linear and nonlinear damping of strength
$\mu > 0$. The nonlinear damping can be described by the nonlinear function

$$f(x_i) = \frac{\gamma}{3}x_i^3, \tag{3.2}$$

where γ is the anisochronicity parameter and $\dot{f}(x_i) = \gamma x_i^2 \dot{x}_i$. The coupling delay is τ,
and the delay-dependent coupling strength in Eq. (3.1) is $\kappa(\tau)$. For the clarity of the
calculations we keep the coupling strength constant in the first part of this Chapter
($\kappa(\tau) = \kappa$), but all analytical results hold for general $\kappa(\tau)$. Since synchronization is
crucially determined by the frequency difference of the two oscillators, we introduce
the detuning parameter $\Delta \in \mathbb{R}$ by

$$\omega_1^2 = \omega_2^2 + \mu\Delta. \tag{3.3}$$

In Fig. 3.3, we display frequency locking as obtained by numerical simulation
of Eq. (3.1) with symmetric initial conditions. The observed angular frequencies Ω

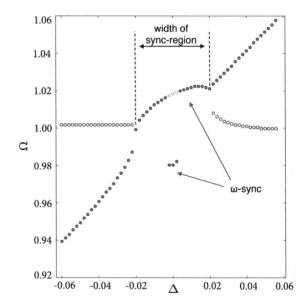

Fig. 3.3 Angular frequency
Ω of oscillator x_1 (dark green
circles), and oscillator x_2
(light yellow circles) versus
the detuning Δ of the
oscillators. Full (empty)
circles correspond to
symmetric (non-symmetric)
initial conditions.
Parameters: $\omega_2 = 1$,
$\mu = 0.1, \gamma = 1, \kappa = 0.4$,
$\tau = 0.1\pi$. Figure from [21]

are plotted versus the detuning Δ of two Van der Pol oscillators. A pronounced synchronization region and a sharp transition to synchronization is observed. Within the synchronization region, only for small $|\Delta|$ the in-phase synchronized solution is observed (lower branch), while for larger $|\Delta|$ the anti-phase synchronized solution (upper branch) is obtained. This solution arises even in the case of symmetric initial conditions (full circles). Note that for small $|\Delta|$, the anti-phase synchronized solution is also observed for any non-symmetric initial conditions (empty circles). Our goal is to analyze the synchronization frequency, the width of the synchronization region, the phase difference in the synchronized state, its stability, and the bifurcation scenarios which occur at the boundaries.

3.3 Analytic Approaches

Dealing with nonlinear systems, it is often necessary to use numerical methods. However in this Section, we will present two analytic approaches which can be employed to the problem at hand.

3.3.1 Method of Averaging

The method of averaging (quasiharmonic reduction) describes weakly nonlinear oscillations in terms of slowly varying amplitude and phase.

For $\mu = 0$ the uncoupled system reduces to the harmonic oscillator $\ddot{x}_i + \omega_i^2 x_i = 0$ with solution

$$x_i = R_i \sin(\omega_i t + \phi_i), \tag{3.4}$$

with constant amplitude R_i and phase ϕ_i. For $0 < \mu \ll 1$ we look for a solution in the form Eq. (3.4) but assume that the amplitude $R_i \geq 0$ and the phase ϕ_i are time-dependent functions:

$$\begin{aligned}
x_i &= R_i(t) \sin(\omega_i t + \phi_i(t)), \\
\dot{x}_i &= R_i(t)\omega_i \cos(\omega_i t + \phi_i(t)),
\end{aligned} \tag{3.5}$$

where terms involving the slowly varying functions $\dot{R}_i$, $\dot{\phi}_i$ are neglected. Without loss of generality, we choose $\omega_2 = 1$. For small μ we use the method of averaging, assuming that the product $\mu\tau$ is small, and Taylor expand $R_i(t - \tau)$ and $\phi_i(t - \tau)$ in the following way:

$$R_i(t - \tau) = R_i(t) - \tau\dot{R}_i(t) + \frac{\tau^2}{2}\ddot{R}_i(t) + \dots . \tag{3.6}$$

We introduce the phase difference $\psi(t) = \phi_1(t) - \phi_2(t)$. Defining a new time scale $\tilde{t} = \frac{2t}{\mu}$, we find the equations which describe the system (3.1) on a slow time scale:

$$\dot{R}_{1/2}(\tilde{t}) = R_{1/2}(\tilde{t}) \left(1 - \frac{\gamma R_{1/2}(\tilde{t})^2}{4}\right) \mp \kappa R_{2/1}(\tilde{t}) \sin(\psi(\tilde{t}) + \tau), \qquad (3.7)$$

$$\dot{\psi}(\tilde{t}) = -\Delta + \kappa \left[\frac{R_1(\tilde{t})}{R_2(\tilde{t})} \cos(\psi(\tilde{t}) - \tau) - \frac{R_2(\tilde{t})}{R_1(\tilde{t})} \cos(\psi(\tilde{t}) + \tau)\right]. \qquad (3.8)$$

For the sake of simplicity, we omit the tilda $\sim$ in the following. The combined effect of the method of averaging and the truncation of the Taylor expansion in τ, is a reduction of the infinite-dimensional problem to a finite-dimensional one (valid only if the product $\mu\tau$ is small). This key step enables us to handle the original delay differential equation as a system of ordinary differential equations [22, 25]. We now have two dynamical equations (3.7) for the amplitudes R_1 and R_2, and one equation (3.8) for the phase difference ψ, which is also called the *slow phase*. The latter equation is a generalized Adler equation [3] and contains the main features of synchronization.

3.3.2 Generalized Adler Equation

The equilibria of the Adler equation correspond to the locking of phase and frequency, since the difference between the phases is constant. To investigate the stability and bifurcation scenario of such fixed points we take a closer look at the generalized Adler equation (3.8), written in general form:

$$\dot{\psi}(t) = -\Delta + \kappa q(\psi(t)), \qquad (3.9)$$

where the averaged forcing term $q(\psi(t))$ is the 2π-periodic function

$$q(\psi(t)) = \frac{R_1(t)}{R_2(t)} \cos[\psi(t) - \tau] - \frac{R_2(t)}{R_1(t)} \cos[\psi(t) + \tau]. \qquad (3.10)$$

The generalized Adler equation (3.8) is a valuable tool for the calculation of the Arnold tongue (see below), which is one of the main characteristics of synchronization in nonlinear systems. For further analysis it is useful to eliminate the amplitudes $R_i(t)$ from Eq. (3.10). Therefore we express $R_2(t)$ by $R_1(t)$ in the case of a relative equilibrium of the amplitudes ($\dot{R}_i(t) = 0$). We achieve two relevant solutions for the stationary amplitude. Inserting the values of R_i into Eq. (3.10) – according to the numerical results – we can plot $\dot{\psi}$ versus ψ in Fig. 3.4a, which gives a graph of the right-hand side of the generalized Adler equation (3.9), i.e., the function $q(\psi)$ in the case $\Delta = 0$.

Fig. 3.4   a The right-hand side of the Adler equation (3.9) $\dot{\psi} = q(\psi)$ for zero frequency detuning $\Delta = 0$ versus the slow phase ψ. The difference between the maximum and minimum gives the width of the synchronization tongue. Parameters: $\omega_1 = \omega_2 = 1$, $\mu = 0.1, \gamma = 1, \kappa = 0.2$, $\tau = 0.1\pi$. **b** Analytic (line) and numeric (dots) results for the width of the synchronization tongue as a function of the delay time τ for $\omega_2 = 1, \mu = 0.1, \gamma = 1$, $\kappa = 0.4$. Figure from [21]

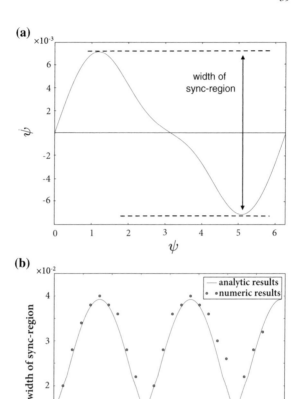

Since in the synchronization region $\dot{\psi} = 0$ and hence $\Delta = q(\psi)$, the maximum and minimum in Fig. 3.4a correspond to the border of the synchronization tongue when varying Δ as we can see in Fig. 3.3. A change of Δ shifts, according to Eq. (3.9), the curve in the y-direction but does not change its shape. In this way we can calculate the width of the synchronization tongue as a function of τ and compare these analytic results to our numerical ones from the simulation of Eq. (3.1) (see Fig. 3.4b). The agreement between the results is remarkable, even though there is an unavoidable small deviation because of the limited numerical accuracy of our simulation. The transient times are very large at bifurcation points. Furthermore, we gain information about the stability of the synchronization state from Fig. 3.4a: For $\psi = 0$ (or, equivalently, $\psi = 2\pi$) we have an unstable equilibrium, and for $\psi = \pi$ (anti-phase oscillation) a stable equilibrium, since $\dot{\psi} < 0$ for $\psi > 0$ and $\dot{\psi} > 0$ for $\psi < 0$. This is in accordance with experimental results [7], as discussed in Fig. 3.9a. The experimentally observed decrease of the amplitude at $\Delta = 0$ indicates an anti-phase oscillation [1].

3.3.3 Describing Function Method

The describing function method (also called the method of harmonic balance) uses frequency domain techniques to investigate limit cycle behavior in nonlinear systems, which is typically represented by a block diagram (Fig. 3.5). This engineering method requires an approximation, but nevertheless often gives a reliable prediction of the frequency ω_s and amplitude A of the limit cycle in a nonlinear system.

The condition for the harmonic balance, i.e., the limit cycle, of our system can be formulated in the following way [11, 12]:

$$\underbrace{G_i^{-1}(i\omega_s)}_{LINEAR} + \underbrace{N(A_i)}_{NONLINEAR} = \frac{E_i}{A_i}e^{-i\Delta\phi_i}, \quad i = 1, 2. \tag{3.11}$$

$G_i(i\omega_s)$ is called the transfer function of the ith node and $N(A_i)$ the describing function of the nonlinear control loop, where ω_s is the angular frequency of the limit cycle and A_i its amplitude. $\Delta\phi_i$ is the phase difference between the oscillation of the limit cycle and the reference input, which has the amplitude E_i. Equation (3.11) holds for a static, odd nonlinearity and a transfer function which behaves like a low-pass filter. For the method of harmonic balance we use Eqs. (3.1) and (3.2):

$$\underbrace{\ddot{x}_i + \omega_i^2 x_i}_{LINEAR} - \underbrace{\mu(\dot{x}_i - \dot{f}(x_i))}_{NONLINEAR} = \mu\kappa x_j(t - \tau). \tag{3.12}$$

The Laplace transform $(X_i(i\omega_s) = \mathscr{L}\{x_i(t)\} = \int_0^\infty x_i(t)e^{-i\omega_s t}dt)$ of Eq. (3.12) leads to

$$\underbrace{\frac{\omega_i^2 - i\omega_s\mu - \omega_s^2}{i\omega_s\mu}}_{G_i^{-1}(i\omega_s)} X_i + F(X_i) = -ie^{-i\omega_s\tau}\underbrace{\frac{\kappa}{\omega_s}X_j}_{E_i}, \tag{3.13}$$

where $F(X_i)$ is the Laplace transform of the nonlinear function $f(x_i)$. The transfer function $G(i\omega_s)$ is the characteristic of the linear part of our system and can be read

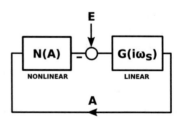

Fig. 3.5 Block diagram of a nonlinear feedback control loop consisting of a linear transfer function $G(i\omega_s)$ and a nonlinear describing function $N(A)$. A stands for the feedback signal and E for the reference input, which is added at the summation point (circle). Figure from [21]

from Eq. (3.13):

$$G_i(i\omega_s) = \frac{i\omega_s\mu}{\omega_i^2 - i\omega_s\mu - \omega_s^2}.\tag{3.14}$$

The ω_s^2-term in the denominator guarantees the low-pass behavior. The describing function $N(A)$ is the amplification of the fundamental harmonics of the periodic signal $x_i(t) = A_i \sin(\omega_s t + \phi_i)$ by the nonlinear function $f(x_i) = \frac{\gamma}{3}x_i^3$ of our system:

$$N(A_i) = \frac{F(X_i)}{X_i} = \frac{f(x_i)}{x_i} \approx \frac{\gamma}{4}A_i^2.\tag{3.15}$$

The approximation of the describing function method is based on the exclusive consideration of the fundamental harmonics in Eq. (3.15). For its calculation, we have used the following trigonometric addition theorem:

$$\sin^3(\zeta) = \frac{1}{4}\left[3\sin(\zeta) - \sin(3\zeta)\right].\tag{3.16}$$

According to the right-hand side of Eq. (3.13), the reference input of one system is given by the delayed output of the other system. The factor $-ie^{-i\omega_s\tau}$ means a negative phase shift of $\frac{\pi}{2} + \omega_s\tau$ in the time domain, so that the reference input in case of synchronization is given by $\frac{\kappa A_j}{\omega_s}\sin(\omega_s(t - \tau) + \phi_j - \frac{\pi}{2})$. In the case of the ith oscillator, the right-hand side of Eq. (3.11) yields

$$\frac{E_i}{A_i}e^{-i\Delta\phi_i} = \frac{\kappa A_j}{\omega_s A_i}e^{-i[\phi_i - \phi_j + \omega_s\tau + \frac{\pi}{2}]}.\tag{3.17}$$

By applying Eqs. (3.14)–(3.17) to Eq. (3.11), we obtain

$$\frac{i}{\mu}\left(\omega_s - \frac{\omega_i^2}{\omega_s}\right) - 1 + \frac{\gamma}{4}A_i^2 = \frac{\kappa A_j}{\omega_s A_i}e^{-i[\phi_i - \phi_j + \omega_s\tau + \frac{\pi}{2}]}.\tag{3.18}$$

The imaginary part of Eq. (3.18) gives us information about the synchronization frequency ω_s versus the time delay. Multiplying it by the imaginary part of the analog equation for the jth oscillator, we obtain:

$$\left(\omega_s - \frac{\omega_i^2}{\omega_s}\right)\left(\omega_s - \frac{\omega_j^2}{\omega_s}\right) = \frac{\mu^2\kappa^2}{\omega_s^2}\cos(\psi + \omega_s\tau)\cos(\psi - \omega_s\tau),\tag{3.19}$$

where we have introduced the phase difference $\psi = \phi_1 - \phi_2$.

The two analytic approaches in this Section yield Eqs. (3.9) and (3.19), respectively. These derived equations reveal information about the phase difference and the frequency of the coupled oscillators, respectively. We discuss these results and their implications in the following Section.

3.4 Occurrence of Synchronization Effects

3.4.1 Arnold Tongue

One important theoretical question is the transition to synchronization, usually characterized in the parameter plane of frequency detuning Δ and coupling strength κ. In the case of delayed coupling the time delay τ has an importance comparable to the coupling strength κ. The synchronization region in the (κ, Δ) or (τ, Δ) plane is generally called Arnold tongue, and it is one of the characteristic features of synchronizing nonlinear systems. First, we calculate the synchronization region analytically using the methods from the previous Section in the plane of the delay time τ and the detuning Δ (Fig. 3.6), in excellent agreement with the numeric results from Eq. (3.1). The boundaries of the Arnold tongue are modulated periodically with a period π as τ is varied. Note that for our choice of $\omega_2 = 1$ the period of the uncoupled harmonic oscillator is 2π. All following calculations hold also in the case of replacing the coupling strength κ by a τ-dependent coupling function $\kappa(\tau)$, see Sect. 3.6.

3.4.2 In- and Anti-phase Mode

Equations (3.7) and (3.8) possess two equilibrium solutions, an in-phase and an anti-phase mode, as we will demonstrate below. The in-phase and anti-phase mode correspond to the enhancement or cancellation of sound in organ pipe experiments, respectively. As we recognize from Fig. 3.3 the center of the synchronization region plays a special role. This motivates a first investigation of the solutions and their stability for vanishing detuning $\Delta = 0$. Such a special parameter setting reduces the technical difficulties and nevertheless allows us to make a qualitative and quantitative

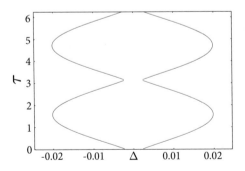

Fig. 3.6 The synchronization region in the parameter plane of delay time τ and frequency detuning Δ. The plot shows an analytically computed Arnold tongue for the coupling strength $\kappa = 0.2$, $\omega_2 = 1, \mu = 0.1, \gamma = 1$. One can see the symmetric, π-periodic boundaries of the tongue. Figure from [21]

analysis of our problem. We find the following equilibrium points for Eqs. (3.7) and (3.8), which, in turn, correspond to the in-phase and anti-phase modes of Eq. (3.1):

$$
\psi_{in} = 0 \Leftrightarrow R_1 = R_2 = \frac{2}{\sqrt{\gamma}}\sqrt{1 - \kappa \sin \tau},
$$

$$
\psi_{anti} = \pi \Leftrightarrow R_1 = R_2 = \frac{2}{\sqrt{\gamma}}\sqrt{1 + \kappa \sin \tau}.
$$

$$(3.20)$$

In order to determine the stability of the in-phase and anti-phase modes, we linearize Eqs. (3.7) and (3.8) around the equilibrium points, which gives the Jacobian matrix J of the system:

$$
J = \begin{pmatrix}
1 - \frac{3}{4}\gamma R_1^2 & -\kappa \sin(\psi + \tau) & -\kappa R_2 \cos(\psi + \tau) \\
\kappa \sin(\psi - \tau) & 1 - \frac{3}{4}\gamma R_2^2 & \kappa R_1 \cos(\psi - \tau) \\
\kappa \left[\frac{R_2 \cos(\psi+\tau)}{R_1^2} + \frac{\cos(\psi-\tau)}{R_2}\right] & -\kappa \left[\frac{\cos(\psi+\tau)}{R_1} + \frac{R_1 \cos(\psi-\tau)}{R_2^2}\right] & \kappa \left[\frac{R_2 \sin(\psi+\tau)}{R_1} - \frac{R_1 \sin(\psi-\tau)}{R_2}\right]
\end{pmatrix}.
$$

$$(3.21)$$

The eigenvalues λ_i, $i = 1, 2, 3$ of the Jacobian matrix evaluated at these equilibrium points determine their linear stability. They are calculated from the characteristic equation

$$
\det(J - \lambda_i \mathbb{1}) = 0. \tag{3.22}
$$

in dependence on the system parameters γ, κ, and τ. Let us first consider the in-phase mode $\psi_{in} = 0$ in Eq. (3.20):

$$
(\lambda + 2 - 2\kappa \sin \tau)[6\kappa^2 + \lambda(\lambda + 2) - 2\kappa \, (\kappa \cos 2\tau + (3\lambda + 2) \sin \tau)] = 0. \tag{3.23}
$$

This equation depends upon the two parameters κ and τ. The boundary of stability with respect to saddle-node bifurcations is given by the condition $\lambda_1 = 0$, which defines the generic saddle-node bifurcation curves in the (κ, τ) plane:

$$
\kappa \left[\kappa - \sin \tau \left(1 - 2\kappa \sin \tau + \kappa^2(1 + \sin^2 \tau)\right)\right] = 0. \tag{3.24}
$$

We obtain three solution branches for κ fulfilling Eq. (3.24):

$$
\kappa_1 = 0,
$$

$$
\kappa_2 = \frac{1}{\sin \tau},
$$

$$
\kappa_3 = \frac{\sin \tau}{1 + \sin^2 \tau}.
$$

$$(3.25)$$

In the case of the anti-phase mode $\psi_{anti} = \pi$ in Eq. (3.20) the characteristic equation (3.22) reads:

$$(\lambda + 2 + 2\kappa \sin \tau)[6\kappa^2 + \lambda(\lambda + 2) - 2\kappa (\kappa \cos 2\tau - (3\lambda + 2) \sin \tau)] = 0.$$
(3.26)

The generic saddle-node bifurcation curves in the (κ, τ) plane is given by:

$$\kappa_1 = 0,$$
$$\kappa_2 = -\frac{1}{\sin \tau},$$
$$\kappa_3 = -\frac{\sin \tau}{1 + \sin^2 \tau}.$$
(3.27)

Note that $\kappa_1 = 0$ represents an uncoupled system. The bifurcation curves $\kappa_{2/3}$ in Eqs. (3.25) and (3.27) separate the regions of stable and unstable equilibrium in the (κ, τ) plane for in-phase and anti-phase mode, respectively; they are represented in Fig. 3.7a, b, respectively. By fixing a value of κ, e.g., $\kappa = 0.4$ (horizontal dash-dotted line), one can trace the change of stability as τ is changed. For $\tau = 0$, Eqs. (3.23) and (3.26) reduce to

$$\lambda^3 + 4\lambda^2 + 4\lambda(\kappa^2 + 1) + 8\kappa^2 = 0$$
(3.28)

which has no solution λ with positive real part, hence both equilibria are stable for $\tau = 0$. The horizontal dash-dotted line $\kappa = 0.4$ intersects with the bifurcation curve $\kappa_3 = \pm\frac{\sin \tau}{1+\sin^2 \tau}$ as shown in Fig. 3.7a, and hence for $\frac{1}{6}\pi < \tau < \frac{5}{6}\pi$ the in-phase mode becomes unstable (Re $\lambda_i > 0$), whereas the anti-phase mode becomes unstable for $\frac{7}{6}\pi < \tau < \frac{11}{6}\pi$, see Fig. 3.7b. In the remaining ranges of τ the in-phase and anti-phase modes, respectively, are stable. Note that bistability of in-phase and anti-phase mode occur around $\tau = 0$ and $\tau = \pi$, as also visible in Fig. 3.3 for $\tau = 0.1\pi$. In Fig. 3.4a, a smaller value $\kappa = 0.2$ is chosen, and hence for $\tau = 0.1\pi$ the in-phase mode is unstable and no bistability exists, in full agreement with Fig. 3.7.

3.4.3 Synchronization Frequency

We recall that the describing function method yields Eq. (3.19) which determines the synchronization frequency ω_s, if the phase difference ψ is known. The method of averaging on the other hand yields the generalized Adler equation (3.9) determining the dynamics of ψ. In Fig. 3.4a, we have found numerically with the help of the Adler equation a stable ($\psi = \pi$) and an unstable ($\psi = 0$) equilibrium point of ψ.

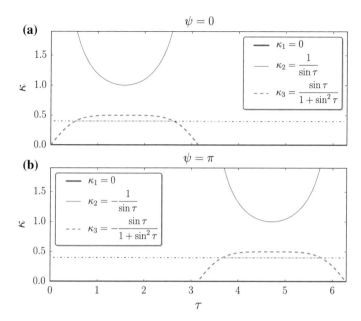

Fig. 3.7 The saddle-node bifurcation curves ($\lambda_1 = 0$) of the characteristic equation (3.22) for $\Delta = 0$ in the plane of coupling strength κ and delay time τ given by Eq. (3.25) for the in-phase mode (**a**) and Eq. (3.27) for the anti-phase mode (**b**). The horizontal dashed-dotted line represents $\kappa = 0.4$ and its grey shaded part is the unstable region with Re $\lambda_i > 0$. The other parameters are $\omega_1 = \omega_2 = 1$, $\mu = 0.1$, $\gamma = 1$. Figure from [21]

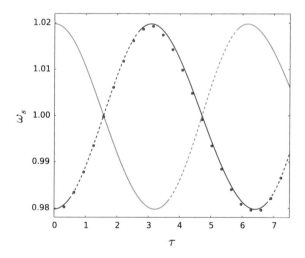

Fig. 3.8 Comparison of the analytic (line) and numeric (dots) results of the synchronization frequency ω_s versus the delay time τ for $\Delta = 0$, $\omega_1 = \omega_2 = 1$, $\mu = 0.1$, $\gamma = 1$, $\kappa = 0.4$. The analytic solution gives an in-phase (dark blue line) and an anti-phase mode (light green line), while the numeric solution (dots) with symmetric initial conditions only reproduces the in-phase mode. Solid line means stable solution, whereas dashed line stands for an unstable one as shown in Fig. 3.7. Figure from [21]

In order to compare the numerical simulations with the results of the describing function method, we set $\omega_1 = \omega_2 = 1$, i.e., $\Delta = 0$. In this case the equilibrium solutions of ψ are given by Eq. (3.20), and hence Eq. (3.19) can be simplified to

$$\omega_s^2 = 1 \pm \mu\kappa \cos(\omega_s\tau) \tag{3.29}$$

where $+$ and $-$ correspond to anti-phase and in-phase oscillations, respectively. In Fig. 3.8, we plot the synchronization frequency ω_s versus the time delay τ for $\Delta = 0$. The congruence between the numerical (from Eq. (3.1)) and analytical result for the in-phase mode (dark blue line, from Eq. (3.29)) is excellent. It is remarkable that the synchronization frequency ω_s is modulated around the single oscillator frequencies $\omega_i = 1$ in dependence upon the delay time. For small delay time, for instance, the in-phase oscillation frequency is lowered, while the anti-phase oscillation frequency (light green line) is increased. The stability of the two branches changes as τ is varied, as discussed above (see Fig. 3.7): At the extrema of the frequency curve in Fig. 3.8 we find bistability.

Note that for symmetric initial conditions the in-phase mode is found as numerical solution for all delay times, although it is in fact unstable in some parts of the τ range (but in these parts, the anti-phase mode would be found for any non-symmetric initial conditions). In Fig. 3.3, the upper frequency branch in the synchronization region stays in the stable anti-phase mode for non-zero detuning Δ (in congruence with experimental data [1]), whereas the lower branch, i.e., the stable in-phase mode, which is close to its instability point, is only observable in a small range of Δ.

3.5   Comparison with Acoustic Experiments

A comparison of a complex experiment with a simple oscillator model is an ambitious endeavor: On the one hand, there is an organ pipe with a whole spectrum of overtones and a complicated aeroacoustic behavior, while, on the other hand, we only consider a simple Van der Pol oscillator. Nevertheless, such a simple model can indeed already exhibit complicated dynamical scenarios, as demonstrated above. Moreover, these scenarios qualitatively agree well with the experiment, as we show by comparing the graphs below. Consequently, our model in fact provides a profound comprehension of the dynamical behavior observed in organ pipes. This supports our point of view that complex behavior can emerge from simple systems, as it is generally accepted in dynamical systems theory.

For visual comparison of experiment and theory, we show the synchronization region versus the detuning frequency of the two oscillators in Fig. 3.9. Both plots show very similar features. Especially the behavior of the transition regions at the two boundaries of the locking interval is remarkable, as well as the concave curvature of the synchronization region itself.

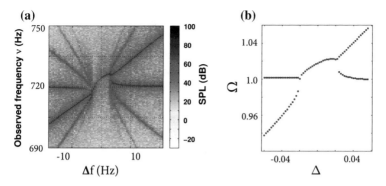

Fig. 3.9 Comparison of synchronization region in experiment and theory: **a** experimentally observed sound pressure level (SPL) in the plane of observed frequency ν versus frequency detuning Δf (in Hz) [7] and **b** numerically calculated angular frequency Ω versus dimensionless detuning Δ for $\omega_2 = 1, \mu = 0.1, \gamma = 1, \kappa = 0.4, \tau = 1.1\pi$. Figure from [21]

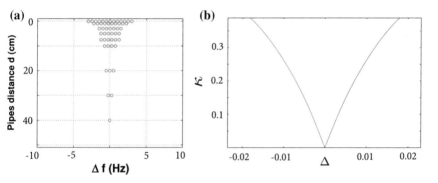

Fig. 3.10 Comparison of experiment and theory: Arnold tongue in the plane of coupling strength κ versus detuning Δ: **a** experiment [7], where the coupling strength is given by the distance d between the organ pipes, and **b** analytic result for $\omega_2 = 1, \mu = 0.1, \gamma = 1$, and delay time $\tau = 0.1\pi$. Figure from [21]

In Fig. 3.10, we compare the experimentally observed and the analytically calculated Arnold tongue in the plane of the coupling strength κ and the detuning Δ. Experimentally, the coupling strength is determined by the distance of the two organ pipes. The analytical calculation proceeds as described in Sect. 3.3. As a result we obtain an Arnold tongue with nonlinear, curved boundaries, see Fig. 3.10b. This is a noticable result which occurs already for a small delay time τ and agrees well with experiments [7, 8], see Fig. 3.10a. The curvature of the boundaries may be further adjusted in our calculations by replacing the constant κ by a τ-dependent coupling strength $\kappa(\tau)$ in the next Section.

3.6   Delay-Dependent Coupling Strength

For a more refined modeling of the non-monotonic behavior of the Arnold tongue and its strong widening at small delay times τ (as observed in experiment, see Fig. 3.11a for $d = 5$ cm), a coupling strength $\kappa(\tau)$ which depends on the delay time τ, should be considered. Recall that we earlier used a constant coupling coefficient κ, see e.g., Eq. (3.1). The delay in the coupling origins from the finite travel time of the sound along the distance d between the pipes. The coupling strength thus also depends on that distance since the sound wave is attenuated according to the radiation of a spherical wave emitted from the pipe mouth. Within the coupling strength $\kappa(\tau)$ we have a near-field term ($\propto \frac{1}{\tau^2}$) and a far-field term ($\propto \frac{1}{\tau}$) with coefficients $\kappa_n, \kappa_f > 0$:

$$\kappa(\tau) = \frac{\kappa_n}{\tau^2} + \frac{\kappa_f}{\tau}. \tag{3.30}$$

By replacing $\kappa(\tau)$ in Eq. (3.1) by Eq. (3.30) we are able to model the boundaries of the Arnold tongue as well as its widening for small delay time τ more realistically as shown in Fig. 3.11b. The reason for the widening is the strong increase of the coupling for small τ in Eq. (3.30):

$$\lim_{\tau \to 0} \kappa(\tau) = \infty. \tag{3.31}$$

Another consequence of introducing the delay-dependent coupling strength is that the in-phase solution in Eq. (3.24) becomes unstable for $\tau = 0$, because κ_2 in Eq. (3.25) (see red curve in Fig. 3.12a) tends to zero. In the case of $\kappa_n = \kappa_f = \kappa$ for example $\kappa_2(0)$ reads

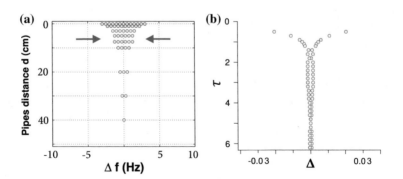

Fig. 3.11 Comparison of experiment and theory: Arnold tongue in the plane of coupling strength κ versus detuning Δ: **a** experiment [7], where the coupling strength is given by the distance d between the organ pipes, and **b** numerical result of Eqs. (3.1) and (3.30) with $\omega_2 = 1$, $\mu = 0.1$, $\gamma = 1$, $\kappa_n = \kappa_f = 0.04$. The red arrows in (**a**) indicate the non-monotonic behavior of the Arnold tongue. Figure from [21]

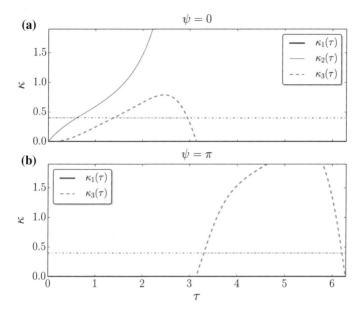

Fig. 3.12 The saddle-node bifurcation curves as in Fig. 3.7 for the case of a delay-dependent coupling strength $\kappa(\tau)$ given by Eq. (3.30) for $\kappa_n = \kappa_f = \kappa$ in the plane of coupling strength κ and delay time τ given by Eq. (3.25) for the in-phase mode (**a**) and Eq. (3.27) for the anti-phase mode (**b**). The horizontal dash-dotted line represents $\kappa = 0.4$ and its grey shaded part is the unstable region with Re $\lambda_i > 0$. The other parameters are as in Fig. 3.7. Figure from [21]

$$\lim_{\tau \to 0} \kappa_2(\tau) = \frac{\tau^2}{(1 + \tau) \sin \tau} = 0. \tag{3.32}$$

In this case just the anti-phase mode is stable. As we can see in Fig. 3.12 compared to Fig. 3.7 the introduction of a delay-dependent coupling strength $\kappa(\tau)$ distorts the bifurcation curves.

3.7 Summary

In this Chapter, we have investigated the synchronization of organ pipes using the tools of nonlinear dynamics. Particular attention has been paid to the delay in the coupling which naturally occurs due to a finite distance of pipes. We have used a simplified nonlinear oscillator model for the pipes, i.e., two coupled Van der Pol oscillators which interact by a dissipative, direct, and delayed coupling. To understand theoretically the dynamics of the system, we have analyzed the locking scenarios. Furthermore, we have numerically integrated the system, and have found that the solution agrees well with existing experiments. On this basis, we have systematically varied the coupling parameters, namely the coupling strength κ and the coupling delay τ.

For a deeper understanding of the various bifurcation scenarios we have developed and extended two complementary analytical approaches: By the method of averaging we obtain a generalized Adler equation for the phase dynamics, which allows us to study the stability of the equilibria corresponding to frequency locking of the oscillators. However, the averaging method does not provide information about the frequency in the locking region. The frequency, in contrast, can be found by the describing function method which allows us to determine the synchronization frequency and hence explain the curvature of the frequency versus detuning. The latter is found in the numerical simulations as well as in the experiments (see Fig. 3.9). Altogether these approximations provide a detailed and complete analytic picture of both relative phase and frequency. In general we obtain excellent agreement of our analytic results with the numerical simulations and with experiments.

A detailed bifurcation analysis has affirmed the existence of in- and anti-phase synchronization. In each case the synchronization frequency has a different value which is in perfect accordance with our analytic calculations. The behavior of the boundaries of the Arnold tongue in the plane of coupling strength κ and detuning Δ depends on the interplay of the coupling strength and the coupling delay time τ. In general, the nonlinear interdependence of κ and τ leads to curved boundaries in the (κ, Δ) plane, which is also clearly confirmed by experimental data. By introducing a delay-dependent coupling strength $\kappa(\tau)$ we can explain the details of the experimentally observed Arnold tongue.

It is interesting to note that there is some similarity of our delayed coupling with the viscoelastic coupling, which has been used in a recent study of two modified Van der Pol oscillators with the aim to describe cardiac synchronization [24]. This viscoelastic coupling is modeled within the Maxwell model of viscous creep by an additional differential equation describing a harmonic spring in series with a linear damper of damping rate (viscosity) γ. This linear inhomogeneous differential equation can be eliminated using a Green's function approach, thereby introducing a distributed delay in the coupling of the two oscillators with an exponential delay kernel with decay rate γ corresponding to a temporal memory [14–16]. In this viscoelastic model also in-phase and anti-phase synchronization scenarios have been found.

References

1. Abel M, Bergweiler S, Gerhard-Multhaupt R (2006) Synchronization of organ pipes: experimental observations and modeling. J Acoust Soc Am **119**:2467–2475
2. Abel M, Ahnert K, Bergweiler S (2009) Synchronization of sound sources. Phys Rev Lett **103**:114301
3. Adler R (1973) A study of locking phenomena in oscillators. Proc IEEE **61**:1380–1385
4. Bader R (2013) Nonlinearities and synchronization in musical acoustics and music psychology. Springer, Berlin
5. Bergweiler S (2006) Körperoszillation und Schallabstrahlung akustischer Wellenleiter unter Berücksichtigung von Wandungseinflüssen und Kopplungseffekten: Verändern Metalllegierung und Wandungsprofil des Rohrresonators den Klang der labialen Orgelpfeife? Ph.D. thesis, Universität Potsdam

6. Fabre B, Hirschberg A (2000) Physical modeling of flue instruments: a review of lumped models. Acta Acust **86**:599
7. Fischer JL (2014) Nichtlineare Kopplungsmechanismen akustischer Oszillatoren am Beispiel der Synchronisation von Orgelpfeifen. Ph.D. thesis, Universität Potsdam
8. Fischer JL, Bader R, Abel M (2016) Aeroacoustical coupling and synchronization of organ pipes. J Acoust Soc Am **140**:2344
9. Fletcher NH (1978) Mode locking in nonlinearly excited inharmonic musical oscillators. J Acoust Soc Am **64**:1566
10. Flunkert V, Fischer I, Schöll E (2013) Dynamics, control and information in delay-coupled systems. Theme Issue of Phil Trans R Soc A **371**:20120465
11. Föllinger O (1993) Nichtlineare Regelungen 2: Harmonische Balance, Popow- und Kreiskriterium, Hyperstabilität, Synthese im Zustandsraum: mit 18 Übungsaufgaben mit Lösungen. De Gruyter
12. Ghoshal G, Chi L, Barabási AL (2013) Uncovering the role of elementary processes in network evolution. Sci Rep **3**:2920
13. Howe MS (2003) Theory of vortex sound, vol 33. Cambridge University Press, Cambridge
14. Kyrychko YN, Blyuss KB, Schöll E (2011) Amplitude death in systems of coupled oscillators with distributed-delay coupling. Eur Phys J B **84**:307–315
15. Kyrychko YN, Blyuss KB, Schöll E (2013) Amplitude and phase dynamics in oscillators with distributed-delay coupling. Phil Trans R Soc A **371**:20120466
16. Kyrychko YN, Blyuss KB, Schöll E (2014) Synchronization of networks of oscillators with distributed-delay coupling. Chaos **24**:043117
17. Pikovsky A, Rosenblum MG, Kurths J (2001) Synchronization: a universal concept in nonlinear sciences. Cambridge University Press, Cambridge
18. Rayleigh JWS (1882) On the pitch of organ-pipes. Philos Mag **13**:340
19. Sawicki J (2015) Synchronization of organ pipes. Master's thesis, Technische Universität Berlin
20. Sawicki J, Abel M, Schöll E (2015) Synchronization in coupled organ pipes. In: Proceedings of the 7th international conference on physics and control (PhysCon (2015) edited by (IPACS Electronic Library, 2015). Istanbul, Turkey
21. Sawicki J, Abel M, Schöll E (2018) Synchronization of organ pipes. Eur Phys J B **91**:24
22. Semenov V, Feoktistov A, Vadivasova T, Schöll E, Zakharova A (2015) Time-delayed feedback control of coherence resonance near subcritical Hopf bifurcation: theory versus experiment. Chaos **25**:033111
23. Stanzial D, Bonsi D, Gonzales D (2001) Nonlinear modelling of the mitnahme-effekt in coupled organ pipes. In: International symposium musical acoustics, vol 108, pp 333
24. Stein S, Luther S, Parlitz U (2017) Impact of viscoelastic coupling on the synchronization of symmetric and asymmetric self-sustained oscillators. New J Phys **19**:063040
25. Wirkus S, Rand RH (2002) The dynamics of two coupled van der pol oscillators with delay coupling. Nonlinear Dyn **30**:205

Chapter 4
Chimeras in Networks Without Delay

In the previous Chap. 3, we have discussed the synchronization mechanism in a system of two coupled oscillators. In the case of two nodes either there exists synchronization or there exists desynchronization. Dealing with systems of more than two nodes gives rise to scenarios between synchronization and desynchronization. We call these states partial synchronization and a well-known prototype of this dynamical scenario are chimera states. Chimera states are an example of intriguing partial synchronization patterns appearing in networks of identical oscillators with symmetric coupling scheme. They exhibit a hybrid structure combining coexisting spatial domains of coherent (synchronized) and incoherent (desynchronized) dynamics, and have been first reported for the model of phase oscillators [1, 30]. Recent studies have demonstrated the emergence of chimera states in a variety of topologies, and for different types of individual dynamics.

The aim of this Chapter is to study chimera states in a network of non-locally coupled Stuart-Landau oscillators. Motivated by former studies, we discuss how a specific set of initial conditions initially separating the network into distinct domains gives rise to a clustered chimera state. Furthermore, the interplay between these initial conditions and non-local coupling is studied. Considering the dynamics of chimera states, our argument shows how "flipped" profiles of the mean phase velocities can be explained by a change of sign of the coupling phase. By this, one can either choose a concave ("upside") profile of the mean phase velocities, or a "flipped" one. These profiles are believed to be a distinct feature of (phase) chimeras, at least in the case of non-locally coupled systems. Extending our reasoning, we show that this argument intuitively explains the transition from phase- to amplitude-mediated chimera state as a result of increasing coupling strength.

The structure of the present Chapter, which includes contents that have been published in [25, 26], is as follows (reproduced content from [26] with the permission of American Institute of Physics (AIP) Publishing): In Sect. 4.1, we introduce the phenomenon of chimera states as a special case of partial synchronization, where incoherent and coherent domains coexist in space. We give an example for a chimera state in a network with a non-local topology in Sect. 4.2. More specifically, we

© Springer Nature Switzerland AG 2019
J. Sawicki, *Delay Controlled Partial Synchronization in Complex Networks*,
Springer Theses, https://doi.org/10.1007/978-3-030-34076-6_4

investigate the impact and relevance of initial conditions for such spatio-temporal patterns in Sect. 4.3. Based on an analytical argument, we show how the coupling phase and the coupling strength are linked to the occurrence of chimera states, flipped profiles of the mean phase velocity, and the transition from a phase- to an amplitude-mediated chimera state. In Sect. 4.4, we summarize the results of this Chapter.

4.1 Chimera States

Chimera states are peculiar partial synchronization patterns that refer to a hybrid dynamics where coherence and incoherence emerge simultaneously in a network of identical oscillators. Since their inception [1, 30], chimera states have attracted massive interest from the nonlinear community for both their significance in understanding complex spatiotemporal patterns and their probable applicabilities in various fields, especially in neuroscience [43]. The last decade has seen an increasing interest in chimera states in dynamical networks [34, 55, 56, 65]. First obtained in systems of phase oscillators [1, 30], chimeras can also be found in a large variety of different systems including time-discrete maps [47, 67, 79], time-continuous chaotic models [48], neural systems [23, 50, 52, 75], Boolean networks [59], population dynamics [5, 24], Van der Pol oscillators [53, 78], and quantum oscillator systems [7]. Moreover, chimera states allow for higher spatial dimensions [42, 49, 56, 73]. Together with the initially reported chimera states, which consist of one coherent and one incoherent domain, new types of these peculiar states having multiple [50, 53, 70, 81, 85] or alternating [22] incoherent regions, as well as amplitude-mediated [71, 72], and pure amplitude chimera and chimera death states [4, 87] have been discovered. A classification has recently been given in [28], where the authors have introduced concepts to distinguish between stationary, turbulent, and breathing chimeras.

In many systems, the form of the coupling defines the possibility to obtain chimera states. The nonlocal coupling has generally been assumed to be a necessary condition for chimera states to evolve in coupled systems. However, recent studies have shown that even global all-to-all coupling [11, 63, 64, 72, 86] and local coupling [36], as well as more complex coupling topologies allow for the existence of chimera states [24, 29, 52, 75, 78]. Furthermore, time-varying network structures can give rise to alternating chimera states [12]. Chimera states have also been shown to be robust against inhomogeneities of the local dynamics and coupling topology [52], as well as against noise [41], or they might even be induced by noise [68, 69].

Possible applications of chimera states in natural and technological systems include the phenomenon of uni-hemispheric sleep [57, 58], bump states in neural systems [33, 61], epileptic seizures [60], power grids [45], or social systems [17]. Many works considering chimera states have mostly been based on numerical results. A deeper bifurcation analysis [35, 51] and even a possibility to control chimera states [8, 54, 74] have been obtained only recently.

The experimental verification of chimera states has been first demonstrated in optical [21] and chemical [46, 77] systems. Further experiments involved mechanical [27, 44], electronic [18, 37], optoelectronic delayed-feedback [38] and electrochemical [62, 82] oscillator systems, Boolean networks [59], and optical combs [80].

4.1.1 Characteristics of Chimera States

The initial paper about chimera states in 2002 has been entitled by "Coexistence of coherence and incoherence in nonlocally coupled phase oscillators" [30]. Since then, this "small work" – dedicated to Prof. Haken from Stuttgart (as written in the acknowledgment of that paper) – has exercised a big impact on the fields of nonlinear dynamics. The mentioned coexistence can be shown by an instantaneous spatial distribution of state variables, or in other words by a snapshot. In Fig. 4.1a, the state variables are given by phases ϕ from a phase reduction of the nonlocally coupled complex Ginzburg-Landau oscillators

$$\partial_t \phi(x, t) = \omega - \int G(x - x') \sin\left(\phi(x, t) - \phi(x', t) + \alpha\right) dx', \quad (4.1)$$

with an exponential coupling kernel

$$G(y) = \frac{\kappa}{2} e^{-\kappa |y|}, \quad (4.2)$$

whereby one part of the ring $30 > k > 170$ is in synchrony, whereas the other part $30 < k < 170$ is desynchronized. For the numerical simulation the spatial continuous variable x has been approximated by a finite number of oscillators $k = 0, \ldots, 200$. That fascinating coexistence of coherence and incoherence is complemented by the fact that the mean of the time derivative of the state variables $\bar{\omega}$ is not constant for all oscillators: Compared to the velocity in the synchronized part, the oscillators in the incoherent part are faster and reveal a typical arc-shaped profile as shown in Fig. 4.1b. Already the two authors have suggested the (Kuramoto) order parameter as an additional measure to quantify chimera states in a network of N oscillators:

$$R(t) = \frac{1}{N} \left| \sum_{k=1}^{N} e^{i\phi_k(t)} \right|, \quad (4.3)$$

where $\phi_k(t)$ is the phase of the kth oscillator at time t. For $R = 1$, the network is synchronized in in-phase, whereas in case of $R \approx 0$, the network is desynchronized. For chimera states, the value of the order parameter lies between these two cases. For a more detailed insight into the transition between coherent and incoherent domain, the local (Kuramoto) order parameter is practicable:

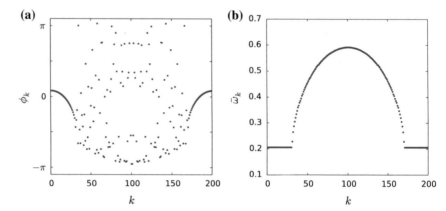

Fig. 4.1 Characteristics of chimera states: **a** Instantaneous spatial distribution of the phases ϕ_k, **b** arc-shaped profile of the mean phase velocities (frequencies) $\bar{\omega}_k$. The continuous Eq. (4.1) is approximated by an array of oscillators $k = 0, \ldots, 200$ with periodic boundary condition. Panels are obtained with initial condition being symmetric about the midpoint $k = 100$ (similar to [30] but with different parameters)

$$R(k, t) = \frac{1}{2P + 1} \left| \sum_{k=N-P}^{N+P} e^{i\phi_k(t)} \right|, \qquad (4.4)$$

where $P \leq N/2$ is a freely selectable number of nodes in an arbitrary neighborhood of the node k. For both measures, we are facing the challenge to define the phase $\phi_k(t)$ in case of a phase-amplitude oscillator, where its dynamics have a specific geometry. For a two dimensional oscillator $\mathbf{x} = (u, v)^T$, we can at best project its limit cycle to the unit circle in the u-v-plane and in this way obtain a geometrical phase $\phi = \arctan(u, v)$. For time-discrete systems, more elaborated procedures have been proposed [47, 84].

Furthermore, recently other correlation measures have been proposed to detect and classify chimeras, e.g., the spatial correlation coefficient g_0 [28], which we will introduce in Chap. 6. For our purpose, it is sufficient to stay mainly with the two measures presented in Fig. 4.1. The characteristic coexistence of spatially separated domains can be shown for instance either by a snapshot as in Fig. 4.1a or a phase-time plot as in Fig. 4.4. The mean phase velocities of the oscillators can be calculated numerically as $\bar{\omega}_k = 2\pi S_k / \Delta T$, where S_k denotes the number of complete rotations realized by the kth oscillator during the time ΔT. However, the chimera state has to be spatially stationary during this time window, otherwise the profile will be smeared out. As will be shown in Chap. 7, in that case special methods can be applied to avoid this problem.

4.1.2 Role of Phase-Lag Parameter

In the following papers about chimera states [1, 2], Abrams and Strogatz have shown
that for a chimera state the phase-lag parameter α in Eq. (4.1) has to be chosen close
to, but smaller than $\pi/2$. In Fig. 4.1, $\alpha = \pi/2 - 0.15 \approx 1.42$ has been taken. Going
beyond simple phase oscillators, we have to translate this condition in an appropriate
way. In Sect. 2.2.1, we have introduced a general equation (2.22) for dynamics on
networks:

$$\dot{\mathbf{x}}_i = \mathbf{f}(\mathbf{x}_i) + \sigma \sum_{j=1}^{N} A_{ij} \mathbf{H}[\mathbf{x}_j - \mathbf{x}_i], \tag{4.5}$$

where $\mathbf{x_i}$ is the local state of node i and $\mathbf{f}$ the local dynamics. The interaction is
realized through diffusive coupling with coupling scheme $\mathbf{H}$ and coupling strength σ,
whereas the network topology is given by the adjacency matrix A_{ij}. The information
about the phase-lag parameter α should be contained in the coupling matrix $\mathbf{H}$. In the
dynamical equation of a two-dimensional system $\mathbf{x}_k^i = (u, v)^T \in \mathbb{R}^2$, $\mathbf{H}$ is a 2×2
matrix:

$$\mathbf{H} = \begin{pmatrix} H_{11} & H_{12} \\ H_{21} & H_{22} \end{pmatrix}. \tag{4.6}$$

In [50], the authors have found the conditions for $\mathbf{H}$ in case of the FitzHugh-Nagumo
model by applying a phase-reduction technique: $\mathbf{H}$ is given by the rotational coupling
matrix

$$\mathbf{H} = \begin{pmatrix} \cos\phi & \sin\phi \\ -\sin\phi & \cos\phi \end{pmatrix}. \tag{4.7}$$

By choosing $\phi \approx \pi/2$, cross couplings between activator (u) and inhibitor (v) variables are realized. In case of the Van der Pol system the analytical calculations in [53]
have required interaction parameters $b_1 = 1$, $b_2 = 0.1$. In terms of these parameters,
the coupling matrix $\mathbf{H}$ can be described in this case as

$$\mathbf{H} = \begin{pmatrix} 0 & 0 \\ b_1 & b_2 \end{pmatrix}, \tag{4.8}$$

where the Van der Pol oscillator is rewritten in the form of a two-dimensional system
(see Sect. 5.2.1). The averaging method as in Sect. 3.3.1 yield

$$\alpha = \arctan\left(\frac{b_1}{b_2}\right) \approx \pi/2 - 0.1. \tag{4.9}$$

For both local dynamics, the FitzHugh-Nagumo and Van der Pol model, the cross
coupling form can be associated to the phase-lag parameter α in the original Kuramoto model (4.1). In Sect. 4.3.2, we will elaborate upon the phase-lag parameter
in case of Stuart-Landau model.

4.2   Ring of Stuart-Landau Oscillators

The paradigmatic model of the Stuart-Landau oscillator which we investigate in this Chapter is the generic expansion of any oscillatory system near a Hopf bifurcation, i.e., a normal form. We have introduced the general form of this model in Sect. 2.1.1 and will use it in the supercritical case (see Eq. (2.8) with $a_1 = 1$ and $a_2 = 0$). Due to this universality, it has been analyzed in numerous works. For instance, the Stuart-Landau model has been used in studies on such intriguing properties as aging [14], the (de-)stabilizing impact of time delays [15, 16], cluster synchronization [13, 39, 66], distributed time delay [3, 31, 32, 83], time-dependent delay [19], or the specific form of response to random perturbations [40, 76]. This Section is focused on a system as in [25, 26] (reproduced content from [26] with the permission of American Institute of Physics (AIP) Publishing).

4.2.1   Non-local Coupling Topology

We consider a ring network of non-locally coupled Stuart-Landau oscillators. An example of a non-local coupling topology is shown in Fig. 4.2b: In the transition scenario between local and global coupling on a ring network, the non-local coupling means, that every node is connected to its P neighboring nodes. The local dynamics is given by the generic expansion (normal form) of an oscillator near a supercritical Hopf bifurcation

$$\dot{z} = (\lambda + i\omega)z - |z|^2 z, \tag{4.10}$$

where $\lambda \in \mathbb{R}$ is the bifurcation parameter, $\omega > 0$ is the frequency of the self-sustained oscillation and $z \in \mathbb{C}$ is the dynamical variable. In the co-rotating frame [20], applying an appropriate scaling of time t, space x, and z,

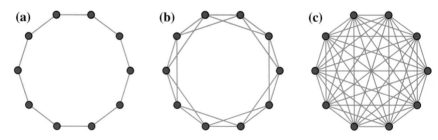

Fig. 4.2 Transition from local (**a**) to global (**c**) coupling topology on a ring via the case of non-local coupling (**b**) for a network of $N = 10$ nodes. In this example each node in the non-local case has $P = 2$ neighbors

$$\tilde{t} = \lambda t,$$
$$\tilde{x} = \lambda^{-1}x,$$ (4.11)
$$\tilde{z} = \lambda^{-1/2}e^{-i\omega t}z,$$

and then dropping the tilde, the local dynamics is simplified to

$$\dot{z} = (1 - |z|^2)z = f(z),$$ (4.12)

where $\lambda > 0$ has been assumed. The network can be described in the continuum limit by the following partial differential equation,

$$\partial_t z(x, t) = f(z) + \sigma e^{i\alpha} \int_0^L G(x - x')\left[z(x') - z(x)\right]dx',$$ (4.13)

where the local dynamics $f(z)$ of an oscillator is given by Eq. (4.12), σ is the coupling strength, α is the coupling phase, L is the system size assuming periodic boundary conditions, and $G(x - x')$ is the coupling kernel determining the functional shape and range of the non-local coupling. Here we assume that the kernel is given by a Gaussian with mean zero

$$G(x - x') = c\,e^{-|x-x'|^2},$$ (4.14)

where $c = 1/\Gamma(\frac{1}{2})$ denotes the normalization factor and Γ is the gamma-function, but our results hold also for more general kernels.

To motivate a specific choice of parameters and initial conditions governing the emergence of chimera states, the system is transformed to polar coordinates via $z = r\exp(i\theta)$. This yields the following partial differential equations that describe the evolution of the amplitude r and phase θ,

$$\partial_t r(x, t) = F(r) + \sigma \underbrace{\int_0^L G(x - x')\left[r(x')\cos(\theta(x') - \theta(x) + \alpha) - r(x)\cos\alpha\right]dx'}_{\Sigma_r},$$
(4.15)

$$\partial_t \theta(x, t) = \sigma \underbrace{\int_0^L G(x - x')\left[\frac{r(x')}{r(x)}\sin(\theta(x') - \theta(x) + \alpha) - \sin\alpha\right]dx'}_{\Sigma_\theta}.$$ (4.16)

The local dynamics of the amplitudes is given by $F(r) = (1 - r^2)r$ with a stable fixed point $r_0 = 1$. In the following we study the impact of the non-local coupling on the dynamics of the network. Introducing the amplitude coupling Σ_r and the phase coupling Σ_θ we can write Eqs. (4.15) and (4.16) as

$$\partial_t r(x, t) = F(r) + \Sigma_r(x, t),$$
$$\partial_t \theta(x, t) = \Sigma_\theta(x, t). \tag{4.17}$$

For the numerical simulations we use the discretized version of Eq. (4.13), i.e., a ring of N coupled oscillators

$$\dot{z}_j = f(z_j) + \sigma e^{i\alpha} \sum_{k=1}^{N} G_{jk} \left[z_k - z_j \right], \tag{4.18}$$

where $j = 1, \dots, N$ and all indices are modulo N. $G_{jk} = \Delta x \, G \, (\Delta x \, [j - k])$ is the discretized version of the coupling kernel in Eq. (4.14), where $\Delta x = L/N$ is the spatial increment between neighboring oscillators.

4.3 The Impact of Initial Conditions

An important issue, often considered as a necessary condition for the existence of chimera states, is the choice of initial conditions. Random initial conditions do not always guarantee chimera behavior. This is due to the fact that classical chimera states typically coexist with the completely synchronized regime. In the case of chimera states the basin of attraction can be relatively small in comparison with that of the synchronized state. In the present Section, we discuss the impact of specially prepared initial conditions and non-local coupling in order to explain, predict and confirm the occurrence of chimera states and their main features. This Section is closely related to [25, 26] (reproduced content from [26] with the permission of American Institute of Physics (AIP) Publishing).

4.3.1 From Initial Conditions to a Clustered Chimera State

Using an anti-phase cluster as initial condition, it is possible to simplify the initial coupling terms in amplitude and phase significantly. The initial conditions are chosen as two clusters in anti-phase,

$$r(x, t_0) = 1,$$
$$\theta(x, t_0) = \begin{cases} \pi, & \text{if } x \in (0, L/2] \\ 0, & \text{if } x \in (L/2, L]. \end{cases} \tag{4.19}$$

The network is initially divided into two equally sized domains. The first one, with phase π, reaches from 0 to $L/2$. The second one, with phase 0, reaches from $L/2$ to L. By this choice of two domains in anti-phase, the network is initially spatially

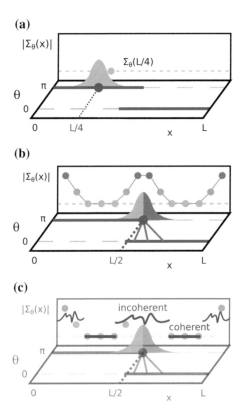

Fig. 4.3 Sketch of the initial dynamical scenario obtained by the choice of two populations initially in anti-phase, as given by Eq. (4.19). The distribution of the phases θ versus space x is shown by full red lines. The coupling kernel $G(x - x')$ localized at a specific oscillator (red dot) is shaded (green). **a** Oscillator at the center of an in-phase population at $x_0 = L/4$, yielding a vanishing coupling term $\Sigma_\theta = 0$. **b** Oscillator at the border between in-phase and anti-phase populations $x_c = L/2$, yielding a maximum coupling term $\Sigma_\theta(x_c)$. The green connected dots sketch the profile of the coupling term Σ_θ versus x. The magnitude of the initial coupling term $\Sigma_\theta(x)$ is illustrated by the brightness of the green color. There are four distinct regions, two where the coupling term nearly vanishes, and two where it does not (note the periodic boundary conditions in x). **c** Sketch of the dynamical scenario arising from this distribution of the initial coupling term. The blue straight lines illustrate coherent states with a constant phase, where the phase dynamics of the oscillators is not perturbed by the coupling term. The red twisted lines denote incoherent states with varying phases. These are centered around the borders between the two oscillator population. In these regions the coupling term does not vanish due to the non-local coupling to oscillators in anti-phase. Reproduced from [26], with the permission of American Institute of Physics (AIP) Publishing

separated into four distinct domains with respect to the coupling terms Σ_r and Σ_θ. This is schematically shown in Fig. 4.3. Two domains, where the coupling terms initially nearly vanish because the oscillators are coupled solely to oscillators in phase (Fig. 4.3a), are separated by two domains where the coupling terms Σ_r and Σ_θ have finite, non-vanishing values due to the coupling to oscillators that are in anti-phase (Fig. 4.3b). This initial separation influences the corresponding long-time behavior significantly. While the dynamics of the two populations with almost vanishing coupling terms becomes synchronized, the two populations where the coupling does not vanish initially, are perturbed in their phase and amplitude dynamics, see Fig. 4.3c. The corresponding chimera state can be clearly seen in a space-time plot, where the dynamics is shown for the real parts $\mathrm{Re}(z_j)$ for every node of the network (Fig. 4.4). The two populations of oscillators being initially in anti-phase split into the four domains mentioned. Two clusters in anti-phase are formed around the centers of the initial in-phase domains at $x = L/4$ ($j = 25$) and $x = 3L/4$ ($j = 75$). The two coherent domains are separated by incoherent domains, their initial centers being at $x = L/2$ ($j = 50$) and $x = L(j = 100)$.

The validity of this approach has been tested for long simulation times and increasing numbers of oscillators forming the network. Our simulations confirm that the observed chimera states are long-living and rule out finite-size effects for oscillator numbers up to $N = 1001$.

4.3.2   Off-Diagonal Coupling Revisited

As we have outlined in Sect. 4.1.2, a phase-lag $\alpha \simeq \pi/2$ is required for chimera states to arise in networks of phase-oscillators [1]. Recently, it has been shown that this holds as well for the coupling phase of FitzHugh-Nagumo oscillators [50]. Also the chimera state shown in Fig. 4.4 requires a coupling phase α close to $\pi/2$ in order to be observed. Such a condition has been used in many studies [1, 50, 74]. We show that the approach outlined in the previous Section gives an intuitive explanation of this property. Furthermore, it allows us to predict and explain the occurrence of "flipped" profiles of the mean phase velocities.

To this purpose, we study the initial dynamics that is simplified by the initial conditions. The amplitude initial conditions, Eq. (4.19), result in vanishing local

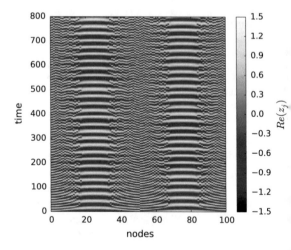

Fig. 4.4 Space-time plot of $\mathrm{Re}(z_j)$ in a network of N non-locally coupled oscillators. The initial conditions are given by Eq. (4.19). Parameters are given by $\sigma = 0.6$, $\alpha = \pi/2 - 0.15$, $N = 101$, $L = 2\pi$. Reproduced from [26], with the permission of American Institute of Physics (AIP) Publishing

dynamics of the amplitudes, $F(r, t_0) = 0$, and the initial dynamics is simplified to

$$\partial_t r(x, t_0) = \Sigma_r(x, t_0),$$
$$\partial_t \theta(x, t_0) = \Sigma_\theta(x, t_0). \tag{4.20}$$

Using the initial conditions, Eq. (4.19), in the definitions of the coupling terms given by Eqs. (4.15) and (4.16), the initial coupling terms are simplified to

$$\Sigma_r(x, t_0) = -\sigma \cos \alpha \, C_r(x),$$
$$\Sigma_\theta(x, t_0) = -\sigma \sin \alpha \, C_\theta(x), \tag{4.21}$$

where the function $C_r(x)$ summarizes the values of the integral in the amplitude dynamics and other constants, and the function $C_\theta(x)$ summarizes the values of the integral in the phase dynamics and other constants. If we now take a look at the scenario sketched in Fig. 4.3, the mechanism leading to a chimera state is uncovered: While the functions representing the integral vanish towards the center of the synchronized domains, leading to synchronized behavior, their non-zero values towards the borders between the anti-phase domains leads to varying, desynchronized behavior.

For α close to $\pi/2$ the amplitude coupling term $\Sigma_r(x, t_0)$ nearly vanishes and the magnitude of the phase coupling term $\Sigma_\theta(x, t_0)$ is maximum, thus effectively restricting the variation to the phases. It is important to note that this effect of the initial coupling terms in Eq. (4.21) also occurs if the coupling phase α approaches the value $-\pi/2$. This property is used in the next Subsection where the occurrence of "flipped" profiles of the mean phase velocities and its connection to the coupling phase α is discussed. The possibility to increase amplitude modulations by a proper choice of coupling strength σ is analyzed in the subsequent Section.

4.3.3   "Flipping" Profiles of the Mean Phase Velocities

From Eq. (4.20) it follows that the sign of the phases is determined by the phase coupling term $\Sigma_\theta(x, t)$ solely. Therefore, a change in the sign of Σ_θ changes the phase dynamics qualitatively. In particular, for positive values of Σ_θ the phases are expected to evolve to positive values while for negative values of Σ_θ the phases become negative. In the first case, a positive phase velocity results in a normal concave "upside" profile of the mean phase velocities $\omega_j = \partial_t \theta(x_j)$, while in the latter case negative values of the phase velocities lead to a convex "flipped" profile of the mean phase velocities, see Fig. 4.5.

The sign of Σ_θ is changed by a suitable choice of α. Coupling phases α close to $-\pi/2$ fulfill the requirement of almost vanishing amplitude coupling terms Σ_r and maximum magnitude of the phase coupling terms Σ_θ, as well. Taking advantage of this, the sign of the coupling terms can be modified by a change of the sign

Fig. 4.5 Snapshots of the phases θ_j (top panels), phase coupling term Σ_θ (middle panels) and profile of the mean phase velocities ω_j (bottom panels) at $t = 400$ for **a** $\alpha = -(\pi/2 - 0.15)$ and **b** $\alpha = \pi/2 - 0.15$. Initial conditions and parameters as in Fig. 4.4. Reproduced from [26], with the permission of American Institute of Physics (AIP) Publishing

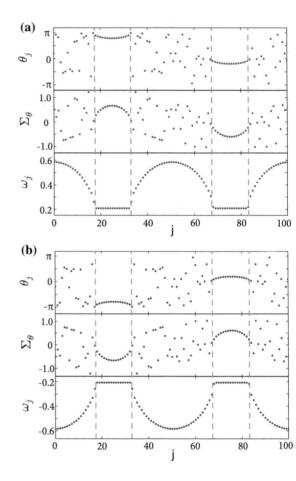

of α. As shown in Fig. 4.5a a value of $\alpha = \pi/2 - 0.15$ leads to a negative sign of the coupling terms Σ_θ, and a "flipped" profile of the mean phase velocities can be observed for the domains of incoherent phases. In contrast, in Fig. 4.5b a choice of $\alpha = -(\pi/2 - 0.15)$ results in a positive coupling term Σ_θ, leading to a normal concave "upside" profile of positive mean phase velocities.

4.3.4   Transition from Phase to Amplitude-Phase Chimera States

A feature of amplitude-mediated chimera states, as reported recently [71], is the coexistence of coherent and incoherent domains not only for the phases but also for the amplitudes. By inspecting the simplified coupling term in the amplitudes, Σ_r,

Fig. 4.6 Snapshots of the phase θ_j (red dots) and amplitudes r_j (blue lines) at $t = 2000$ for **a** weak coupling strength, $\sigma = 0.1$, and **b** increased coupling strength, $\sigma = 0.6$. Initial conditions and other parameters as in Fig. 4.4. Reproduced from [26], with the permission of American Institute of Physics (AIP) Publishing

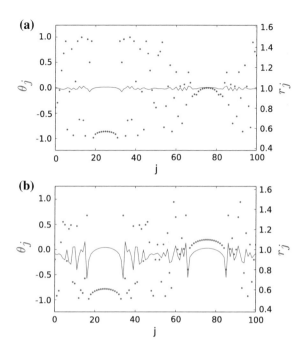

it is possible to explain the transition from phase chimera states to amplitude-phase chimera states by increasing the coupling strength σ. As discussed above, the initial dynamics for the amplitudes is simplified to

$$\partial_t r(x, t_0) = \Sigma_r(x, t_0), \tag{4.22}$$

where the coupling term for the amplitudes is given by

$$\Sigma_r(x, t_0) = -\sigma \cos\alpha \, C_r(x). \tag{4.23}$$

The magnitude of the coupling term Σ_r increases linearly by the coupling strength σ. Therefore, in the limit of weak coupling ($\sigma = 0.1$) the occurrence of a phase chimera is expected, where the variations in the amplitudes are negligible, see Fig. 4.6a.

In contrast, as shown in Fig. 4.6b, for increased values of the coupling strength ($\sigma = 0.6$) the amplitude variations increase and the incoherent dynamics of the phases is combined with non-vanishing modulations in the amplitudes r_j.

4.4 Summary

In the current Chapter, we have investigated partial synchronization in networks. A known example for partial synchronization are chimera states, which combines coexisting spatial domains of synchronized and desynchronized dynamics in a network. In Sect. 4.1, we have given a short introduction and explained the characteristics of chimera states. Moreover, we have outlined the importance of the coupling phase α for these hybrid states. In Sect. 4.2, we have analyzed chimera states in networks of Stuart-Landau oscillators. We have shown that the occurrence of phase chimeras can be seen as caused by a phase-lag in the coupling. In Sect. 4.3, we have provided an analytical argument that explains the need for a pronounced off-diagonal coupling, in order to create chimera states. The provided analytical argument has incorporated the initial conditions and the range of non-local coupling. These two aspects have a great influence on the coupling terms. Recent studies have confirmed the importance of analyzing coupling terms to explain the occurrence of chimera states [9, 10].

Our analytical approach has allowed three statements to be validated: First, we have discussed how this approach provides an intuitive answer to the question why a coupling phase α close to $\pi/2$ (off-diagonal coupling) is needed in order to access chimera states. Second, we have discussed the impact of the sign of the coupling phases. We have explained how a change of the sign of the coupling phase α leads to the occurrence of normal and "flipped" arc-shaped profiles of the mean phase velocities, respectively, and thereby, determines the sign of the profile of the mean phase velocities. Third, we have exemplified how our argument gives an intuitive explanation for the transition from phase chimera states to a coupled phase-amplitude chimera state. We have discussed how an increase of the coupling strength σ is linked to a transition from a pure phase chimera state in the limit of weak coupling to a state sharing the main features of an amplitude-mediated chimera state in the case of intermediate coupling strength. The latter has shown the main properties of an amplitude-mediated chimera state [6, 71], i.e., the variations in the phases are connected with non-vanishing variations in the amplitudes. Our results have been obtained for the paradigmatic model of a network of coupled Stuart-Landau oscillators, and we, therefore, expect their wide applicability.

References

1. Abrams DM, Strogatz SH (2004) Chimera states for coupled oscillators. Phys Rev Lett **93**:174102
2. Abrams DM (2006) Two coupled oscillator models: the Millennium bridge and the Chimera state. Ph.D. thesis, Cornell University
3. Atay FM (2003) Distributed delays facilitate amplitude death of coupled oscillators. Phys Rev Lett **91**:094101
4. Banerjee T (2015) Mean-field-diffusion-induced chimera death state. Europhys Lett **110**:60003
5. Banerjee T, Dutta PS, Zakharova A, Schöll E (2016) Chimera patterns induced by distance-dependent power-law coupling in ecological networks. Phys Rev E **94**:032206

6. Banerjee T, Ghosh D, Biswas D, Schöll E, Zakharova A (2018) Networks of coupled oscillators: from phase to amplitude chimeras. Chaos **28**:113124
7. Bastidas VM, Omelchenko I, Zakharova A, Schöll E, Brandes T (2015) Quantum signatures of chimera states. Phys Rev E **92**:062924
8. Bick C, Martens EA (2015) Controlling chimeras. New J Phys **17**:033030
9. Bogomolov S, Strelkova G, Schöll E, Anishchenko VS (2016) Amplitude and phase chimeras in an ensemble of chaotic oscillators. Tech Phys Lett **42**:765–768
10. Bogomolov S, Slepnev A, Strelkova G, Schöll E, Anishchenko VS (2017) Mechanisms of appearance of amplitude and phase chimera states in a ring of nonlocally coupled chaotic systems. Commun Nonlinear Sci Numer Simul **43**:25
11. Böhm F, Zakharova A, Schöll E, Lüdge K (2015) Amplitude-phase coupling drives chimera states in globally coupled laser networks. Phys Rev E **91**:040901(R)
12. Buscarino A, Frasca M, Gambuzza LV, Hövel P (2015) Chimera states in time-varying complex networks. Phys Rev E **91**:022817
13. Choe CU, Dahms T, Hövel P, Schöll E (2010) Controlling synchrony by delay coupling in networks: from in-phase to splay and cluster states. Phys Rev E **81**:025205(R)
14. Daido H, Nakanishi K (2004) Aging transition and universal scaling in oscillator networks. Phys Rev Lett **93**:104101
15. D'Huys O, Vicente R, Danckaert J, Fischer I (2010) Amplitude and phase effects on the synchronization of delay-coupled oscillators. Chaos **20**:043127
16. Fiedler B, Flunkert V, Georgi M, Hövel P, Schöll E (2007) Refuting the odd number limitation of time-delayed feedback control. Phys Rev Lett **98**:114101
17. Gonzalez-Avella JC, Cosenza MG, Miguel MS (2014) Localized coherence in two interacting populations of social agents. Phys A **399**:24–30
18. Gambuzza LV, Buscarino A, Chessari S, Fortuna L, Meucci R, Frasca M (2014) Experimental investigation of chimera states with quiescent and synchronous domains in coupled electronic oscillators. Phys Rev E **90**:032905
19. Gjurchinovski A, Zakharova A, Schöll E (2014) Amplitude death in oscillator networks with variable-delay coupling. Phys Rev E **89**:032915
20. García-Morales V, Krischer K (2012) The complex Ginzburg-Landau equation: an introduction. Contemp Phys **53**:79–95
21. Hagerstrom AM, Murphy TE, Roy R, Hövel P, Omelchenko I, Schöll E (2012) Experimental observation of chimeras in coupled-map lattices. Nat Phys **8**:658–661
22. Haugland SW, Schmidt L, Krischer K (2015) Self-organized alternating chimera states in oscillatory media. Sci Rep **5**:9883
23. Hizanidis J, Kanas V, Bezerianos A, Bountis T (2014) Chimera states in networks of nonlocally coupled Hindmarsh-Rose neuron models. Int J Bifurcat Chaos **24**:1450030
24. Hizanidis J, Panagakou E, Omelchenko I, Schöll E, Hövel P, Provata A (2015) Chimera states in population dynamics: networks with fragmented and hierarchical connectivities. Phys Rev E **92**:012915
25. Kalle P (2014) Chimera states in Stuart-Landau networks. Master's thesis, Technische Universität Berlin
26. Kalle P, Sawicki J, Zakharova A, Schöll E (2017) Chimera states and the interplay between initial conditions and non-local coupling. Chaos **27**:033110
27. Kapitaniak T, Kuzma P, Wojewoda J, Czolczynski K, Maistrenko Y (2014) Imperfect chimera states for coupled pendula. Sci Rep **4**:6379
28. Kemeth FP, Haugland SW, Schmidt L, Kevrekidis YG, Krischer K (2016) A classification scheme for chimera states. Chaos **26**:094815
29. Ko TW, Ermentrout GB (2008) Partially locked states in coupled oscillators due to inhomogeneous coupling. Phys Rev E **78**:016203
30. Kuramoto Y, Battogtokh D (2002) Coexistence of coherence and incoherence in nonlocally coupled phase oscillators. Nonlinear Phenom Complex Syst **5**:380–385
31. Kyrychko YN, Blyuss KB, Schöll E (2011) Amplitude death in systems of coupled oscillators with distributed-delay coupling. Eur Phys J B **84**:307–315

32. Kyrychko YN, Blyuss KB, Schöll E (2014) Synchronization of networks of oscillators with distributed-delay coupling. Chaos 24:043117
33. Laing CR, Chow CC (2001) Stationary bumps in networks of spiking neurons. Neural Computation 13:1473–1494
34. Laing CR (2009) The dynamics of chimera states in heterogeneous Kuramoto networks. Phys D 238:1569–1588
35. Laing CR (2010) Chimeras in networks of planar oscillators. Phys Rev E 81:066221
36. Laing CR (2015) Chimeras in networks with purely local coupling. Phys Rev E 92:050904(R)
37. Larger L, Penkovsky B, Maistrenko Y (2013) Virtual chimera states for delayed-feedback systems. Phys Rev Lett 111:054103
38. Larger L, Penkovsky B, Maistrenko Y (2015) Laser chimeras as a paradigm for multistable patterns in complex systems. Nat Commun 6:7752
39. Lehnert J, Hövel P, Selivanov AA, Fradkov AL, Schöll E (2014) Controlling cluster synchronization by adapting the topology. Phys Rev E 90:042914
40. Levnajic Z, Pikovsky A (2010) Phase resetting of collective rhythm in ensembles of oscillators. Phys Rev E 82:056202
41. Loos S, Claussen JC, Schöll E, Zakharova A (2016) Chimera patterns under the impact of noise. Phys Rev E 93:012209
42. Maistrenko Y, Sudakov O, Osiv O, Maistrenko VL (2015) Chimera states in three dimensions. New J Phys 17:073037
43. Majhi S, Bera BK, Ghosh D, Perc M (2018) Chimera states in neuronal networks: a review. Phys Life Rev 26
44. Martens EA, Thutupalli S, Fourriere A, Hallatschek O (2013) Chimera states in mechanical oscillator networks. Proc Natl Acad Sci USA 110:10563
45. Motter AE, Myers SA, Anghel M, Nishikawa T (2013) Spontaneous synchrony in power-grid networks. Nat Phys 9:191–197
46. Nkomo S, Tinsley MR, Showalter K (2013) Chimera states in populations of nonlocally coupled chemical oscillators. Phys Rev Lett 110:244102
47. Omelchenko I, Maistrenko Y, Hövel P, Schöll E (2011) Loss of coherence in dynamical networks: spatial chaos and chimera states. Phys Rev Lett 106:234102
48. Omelchenko I, Riemenschneider B, Hövel P, Maistrenko Y, Schöll E (2012) Transition from spatial coherence to incoherence in coupled chaotic systems. Phys Rev E 85:026212
49. Omel'chenko OE, Wolfrum M, Yanchuk S, Maistrenko Y, Sudakov O (2012) Stationary patterns of coherence and incoherence in two-dimensional arrays of non-locally-coupled phase oscillators. Phys Rev E 85:036210
50. Omelchenko I, Omel'chenko OE, Hövel P, Schöll E (2013) When nonlocal coupling between oscillators becomes stronger: patched synchrony or multichimera states. Phys Rev Lett 110:224101
51. Omel'chenko OE (2013) Coherence-incoherence patterns in a ring of non-locally coupled phase oscillators. Nonlinearity 26:2469
52. Omelchenko I, Provata A, Hizanidis J, Schöll E, Hövel P (2015) Robustness of chimera states for coupled FitzHugh-Nagumo oscillators. Phys Rev E 91:022917
53. Omelchenko I, Zakharova A, Hövel P, Siebert J, Schöll E (2015) Nonlinearity of local dynamics promotes multi-chimeras. Chaos 25:083104
54. Omelchenko I, Omel'chenko OE, Zakharova A, Wolfrum M, Schöll E (2016) Tweezers for chimeras in small networks. Phys Rev Lett 116:114101
55. Omel'chenko OE (2018) The mathematics behind chimera states. Nonlinearity 31:R121
56. Panaggio MJ, Abrams DM (2015) Chimera states: Coexistence of coherence and incoherence in networks of coupled oscillators. Nonlinearity 28:R67
57. Rattenborg NC, Amlaner CJ, Lima SL (2000) Behavioral, neurophysiological and evolutionary perspectives on unihemispheric sleep. Neurosci Biobehav Rev 24:817–842
58. Rattenborg NC, Voirin B, Cruz SM, Tisdale R, Dell'Omo G, Lipp HP, Wikelski M, Vyssotski AL (2016) Evidence that birds sleep in mid-flight. Nat Commun 7:12468

59. Rosin DP, Rontani D, Gauthier DJ (2014) Synchronization of coupled Boolean phase oscillators. Phys Rev E **89**:042907
60. Rothkegel A, Lehnertz K (2014) Irregular macroscopic dynamics due to chimera states in small-world networks of pulse-coupled oscillators. New J Phys **16**:055006
61. Sakaguchi H (2006) Instability of synchronized motion in nonlocally coupled neural oscillators. Phys Rev E **73**:031907
62. Schmidt L, Schönleber K, Krischer K, García-Morales V (2014) Coexistence of synchrony and incoherence in oscillatory media under nonlinear global coupling. Chaos **24**:013102
63. Schmidt L, Krischer K (2015) Chimeras in globally coupled oscillatory systems: from ensembles of oscillators to spatially continuous media. Chaos **25**:064401
64. Schmidt L, Krischer K (2015) Clustering as a prerequisite for chimera states in globally coupled systems. Phys Rev Lett **114**:034101
65. Schöll E (2016) Synchronization patterns and chimera states in complex networks: interplay of topology and dynamics. Eur Phys J Spec Top **225**:891–919
66. Selivanov AA, Lehnert J, Dahms T, Hövel P, Fradkov AL, Schöll E (2012) Adaptive synchronization in delay-coupled networks of Stuart-Landau oscillators. Phys Rev E **85**:016201
67. Semenov V, Feoktistov A, Vadivasova T, Schöll E, Zakharova A (2015) Time-delayed feedback control of coherence resonance near subcritical Hopf bifurcation: theory versus experiment. Chaos **25**:033111
68. Semenov V, Zakharova A, Maistrenko Y, Schöll E (2016) Delayed-feedback chimera states: forced multiclusters and stochastic resonance. Europhys Lett **115**:10005
69. Semenova N, Zakharova A, Anishchenko VS, Schöll E (2016) Coherence-resonance chimeras in a network of excitable elements. Phys Rev Lett **117**:014102
70. Sethia GC, Sen A, Atay FM (2008) Clustered chimera states in delay-coupled oscillator systems. Phys Rev Lett **100**:144102
71. Sethia GC, Sen A, Johnston GL (2013) Amplitude-mediated chimera states. Phys Rev E **88**:042917
72. Sethia GC, Sen A (2014) Chimera states: the existence criteria revisited. Phys Rev Lett **112**:144101
73. Shima S, Kuramoto Y (2004) Rotating spiral waves with phase-randomized core in nonlocally coupled oscillators. Phys Rev E **69**:036213
74. Sieber J, Omel'chenko OE, Wolfrum M (2014) Controlling unstable chaos: stabilizing chimera states by feedback. Phys Rev E **112**:054102
75. Tsigkri-DeSmedt ND, Hizanidis J, Hövel P, Provata A (2016) Multi-chimera states and transitions in the leaky integrate-and-fire model with excitatory coupling and hierarchical connectivity. Eur Phys J Spec Top **225**:1149
76. Teramae JN, Tanaka D (2004) Robustness of the noise-induced phase synchronization in a general class of limit cycle oscillators. Phys Rev Lett **93**:204103
77. Tinsley MR, Nkomo S, Showalter K (2012) Chimera and phase cluster states in populations of coupled chemical oscillators. Nat Phys **8**:662–665
78. Ulonska S, Omelchenko I, Zakharova A, Schöll E (2016) Chimera states in networks of Van der Pol oscillators with hierarchical connectivities. Chaos **26**:094825
79. Vadivasova TE, Strelkova G, Bogomolov SA, Anishchenko VS (2016) Correlation analysis of the coherence-incoherence transition in a ring of nonlocally coupled logistic maps. Chaos **26**:093108
80. Viktorov EA, Habruseva T, Hegarty SP, Huyet G, Kelleher B (2014) Coherence and incoherence in an optical comb. Phys Rev Lett **112**:224101
81. Vüllings A, Schöll E, Lindner B (2014) Spectra of delay-coupled heterogeneous noisy nonlinear oscillators. Eur Phys J B **87**:31
82. Wickramasinghe M, Kiss IZ (2013) Spatially organized dynamical states in chemical oscillator networks: synchronization, dynamical differentiation, and chimera patterns. PLoS ONE **8**:e80586
83. Wille C, Lehnert J, Schöll E (2014) Synchronization-desynchronization transitions in complex networks: an interplay of distributed time delay and inhibitory nodes. Phys Rev E **90**:032908

84. Wolfrum M, Omel'chenko OE (2011) Chimera states are chaotic transients. Phys Rev E **84**:015201
85. Xie J, Knobloch E, Kao HC (2014) Multicluster and traveling chimera states in nonlocal phase-coupled oscillators. Phys Rev E **90**:022919
86. Yeldesbay A, Pikovsky A, Rosenblum M (2014) Chimeralike states in an ensemble of globally coupled oscillators. Phys Rev Lett **112**:144103
87. Zakharova A, Kapeller M, Schöll E (2014) Chimera death: symmetry breaking in dynamical networks. Phys Rev Lett **112**:154101

Chapter 5
Interplay of Delay and Fractal Topology

In Chap. 4, we have discussed the emergence of chimera states as an example of intriguing partial synchronization patterns in networks with non-local coupling topology without delay. The topology of the network has been found to play a crucial role in inducing chimera states. While earlier work has focussed on simple nonlocal coupling schemes like rings or two-module structures, chimeras have also been found in all-to-all coupled networks [3, 37, 38, 41, 50], as well as in more complex coupling topologies. Going beyond these regular two-population or nonlocally coupled ring networks, we focus on more complex connectivities, reflecting the structure of real-world networks. Of particular interest are networks with hierarchical connectivities, arising in neuroscience as a result of Diffusion Tensor Magnetic Resonance Imaging analysis, showing that the connectivity of the neuron axons network represents a hierarchical (quasi-fractal) geometry [7, 17–19, 32]. Such a network topology can be realized using a Cantor algorithm starting from a chosen base pattern [5, 26, 45, 51], and is in the focus of the present Chapter, which includes contents that have been published in [35, 36].

Furthermore, we aim to uncover the impact of time delay on the dynamics in such complex network topologies. We analyze the influence of time delay on chimera states in networks with hierarchical connectivity, and demonstrate how by varying the time delay one can stabilize chimera states in the network. In phase oscillators systems, the phase-lag parameter strongly affects the system dynamics and is crucial for the appearance of chimera states. There are two interpretations for the phase-lag parameter [31]: this parameter determines a balance between spontaneous order and permanent disorder [46], or the phase-lag can be interpreted as an approximation for a time-delayed coupling when the delay is small [6]. Larger time delays pertain beyond these possible analogies, and can induce more complex dynamical phenomena. Current analysis of chimera states in oscillatory systems has demonstrated possible ways to control chimera states [2, 28–30, 43], extending their lifetime and fixing their spatial position. It is well known that time delay can also serve as an instrument for stabilization/destabilization of complex patterns in networks.

The intention of this Chapter is as follows: In Sect. 5.1, we introduce networks with hierarchical connectivities. In Sect. 5.2, we elaborate the role of time delay introduced

© Springer Nature Switzerland AG 2019
J. Sawicki, *Delay Controlled Partial Synchronization in Complex Networks*,
Springer Theses, https://doi.org/10.1007/978-3-030-34076-6_5

in the coupling term: In the parameter plane of coupling strength and time delay we find tongue-like regions of existence of chimera states alternating with regions of coherent dynamics, e.g., synchronization and traveling waves. We show analytically and numerically that the period of the synchronized dynamics as a function of delay is characterized by a sequence of piecewise linear branches. In between these branches various chimera states and other partial synchronization patterns are induced by the time delay. By varying the time delay one can deliberately choose and stabilize desired spatio-temporal patterns. Section 5.3 is a conclusion summarizing the results.

5.1   Fractal Topology

We consider a ring of N identical oscillators with different coupling topologies, which are given by the respective adjacency matrix $\mathbf{G}$. While keeping the periodicity of the ring, and the circulant structure of the adjacency matrix, we vary the connectivity pattern of each element. The dynamical equations for the 2-dimensional phase space variable $\mathbf{x}_k = (u_k, v_k)^T \in \mathbb{R}^2$ are:

$$\dot{\mathbf{x}}_k(t) = \mathbf{F}(\mathbf{x}_k(t)) + \frac{\sigma}{g} \sum_{l=1}^{N} G_{kl} \mathbf{H}[\mathbf{x}_l(t - \tau) - \mathbf{x}_k(t)] \tag{5.1}$$

with $k \in \{1, ..., N\}$ and the delay time τ. The parameter σ denotes the coupling strength, and $g = \sum_{l=1}^{N} G_{kl}$ is the number of links for each node (corresponding to the row sum of $\mathbf{G}$). The interaction is realized through diffusive coupling with coupling matrix $\mathbf{H}$. Regarding the dynamics $\mathbf{F}$ of each individual oscillator we will focus on two different paradigmatic models in Sect. 5.2. Beforehand, we want to put the focus on the adjacency matrix $\mathbf{G}$ in this Section.

5.1.1   Fractals and Cantor Construction

In mathematics, fractals are infinitely self-similar, iterated constructs characterized by their fractal dimension d_f, which in contrast to topological dimension is non-integer. Fractal topologies can be generated using a classical Cantor construction algorithm for a fractal set [8, 22]. This iterative hierarchical procedure starts from a *base pattern* or initiation string b_{init} of length b, where each element represents either a link ('1') or a gap ('0'). In Fig. 5.1, such a construction algorithm is represented for the string $b_{init} = (101)$. The number of links contained in b_{init} is referred to as c_1, e.g., in our example $c_1 = 2$. In each iterative step, each link is replaced by the initial base pattern, while each gap is replaced by b gaps. Thus, each iteration increases the size of the final bit pattern, such that after n iterations the total length is $N = b^n$. We call the resulting connectivity fractal or hierarchical, with n as the hierarchical level. Many

Fig. 5.1 Classical Cantor construction algorithm: in each iterative step n, each link (blue, 1) is replaced by the initial base pattern $b_{init} = (101)$, while each gap (empty space) is replaced by $b = 3$ gaps

patterns in nature can be characterized by their "fractal"-like structure. However, in contrast to mathematics, infinite iteration is not possible in nature. Usually, we are faced with self-similar structures to $n = 3 - 5$ levels, e.g., the leaves of ferns, Romanesco broccoli, snow flakes, or blood vessel branching.

Using the string – obtained by the Cantor construction algorithm – as the first row of the adjacency matrix $\mathbf{G}$, we can construct a circulant adjacency matrix $\mathbf{G}$ by applying this string to each element of the ring. In this way, a ring network of $N = b^n$ nodes with hierarchical connectivity is generated [12, 26, 44]. Here, we slightly modify this procedure by including an additional zero in the first instance of the sequence [45], which corresponds to the delayed self-coupling. Therefore, there is no net effect of the diagonal elements of the adjacency matrix G_{kk} on the network dynamics, and hence the first link in the clockwise sense from the reference node is effectively removed from the link pattern. Without our modification, this would lead to a breaking of the base pattern symmetry, i.e., if the base pattern is symmetric, the resulting coupling topology would not be so, since the first link to the right is missing from the final link pattern. As an example we take a closer look at the base pattern (101). The first row of the circulant matrix $\mathbf{G}$ is shown in Table 5.1 for different hierarchical level n.

Our procedure, in contrast, ensures the preservation of an initial symmetry of b_{init} in the final link pattern, which is crucial for the observation of chimera states, since asymmetric coupling leads to a drift of the chimera [2, 28]. Thus, a ring network of $N = b^n + 1$ nodes is generated.

Table 5.1 Fractal topologies generated using a classical Cantor construction algorithm: first row of the circulant adjacency matrix $\mathbf{G}$ for different hierarchical level n for the base pattern $b_{init} = (101)$. A illustration of this algorithm is shown in Fig. 5.1. To ensure the preservation of an initial symmetry a zero is added in the first instance of the sequence. Table taken from [35]

$n = 0$	$G_{1l} = 0101$
$n = 1$	$G_{1l} = 0101000101$
$n = 2$	$G_{1l} = 0101000101000000000101000101$
	$\ldots$

5.1.2 Clustering Coefficient and Effective Coupling Range

Dealing with complex topologies implies the necessity of introducing adequate quantifications for comparison. In the case of a non-local coupling as in Chap. 4 the coupling range $r = P/N$ is sufficient. For its calculation the number of neighbors P together with the system size N is needed. In case of a fractal topology the coupling range can be substituted by the effective coupling range $\bar{r}$ or the link density ρ

$$\bar{r} = \frac{\rho}{2} = \frac{g}{2N}, \tag{5.2}$$

where g is the total number of links of a node. Basically, the effective coupling range $\bar{r}$ gives the normalized information about the total number of links of a node, regardless of whether the links are neighboring or not. Taking into account the fractal aspect of the topology the use of the Hausdorff dimension, lacunarity, or the fractal dimension d_f is reasonable. In case of the circulant matrix $\mathbf{G}$, which has been constructed by the classical Cantor algorithm, the fractal dimension d_f can be simply calculated by

$$d_f = \ln \frac{c_1}{b}, \tag{5.3}$$

where c_1 is the number of links contained in b_{init} and b its length. Former studies [45] introduced a so-called hierarchical step m to increase the effective coupling range, which is decreasing rapidly for higher values of the hierarchal level n. As a consequence of such a sparse network only chaotic dynamics can be observed. The hierarchal step m gives the possibility to tune an appropriate initial base pattern between a non-local ($m = 0$) and a fully fractal topology ($m = n$). A more general measure is the clustering coefficient (see Sect. 2.2.3). There are two definitions of the clustering coefficient, which are both mentioned in [24]:

- Clustering coefficient (transitivity)

$$C = \frac{3 \times \text{number of triangles in the network}}{\text{number of connected triples of nodes}}. \tag{5.4}$$

- Global clustering coefficient (introduced by Watts and Strogatz [47])

$$C' = \frac{1}{N} \sum_k C_k, \tag{5.5}$$

which is given by the mean of the local clustering coefficients

$$C_k = \frac{\text{number of triangles connected to node } k}{\text{number of triples centered on node } k}. \tag{5.6}$$

The definition introduced by Watts and Strogatz [47] in Eq. (5.5) reverses the order of the operations in Eq. (5.4): While the first definition is taking the ratio of

Fig. 5.2 Exemplary network
of four nodes (vertices) and
five links (edges). We obtain
differing values of the
clustering coefficient by
using different definitions:
$C = 0.75$ from Eq. (5.4) and
$C' \approx 0.83$ from Eq. (5.5)

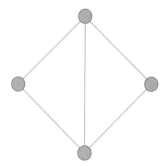

the means, the latter one calculates the mean of the ratio of triangles to triples. By doing so the low-degree nodes are weighted more heavily, because of their small denominator. In the case of a node l with only one neighbor C_l can be defined as either one or zero. For the small network given in Fig. 5.2, we obtain different values of clustering coefficient:

$$C = \frac{3}{4}, \quad C' = \frac{5}{6}. \tag{5.7}$$

Although it is numerically easier to calculate C', one should be aware of the difference: Different computer programs use different definitions without giving the exact one explicitly: For example *Wolfram mathematica* is using the first definition, while the *boost library* the second one. As a further example for the mismatch between both definitions we provide a comparison between the clustering coefficient calculated for different hierarchical steps m. In Fig. 5.3a, Eq. (5.4) is used as the definition, whereas in Fig. 5.3b Eq. (5.5). Particularly for the base pattern (110) there is a big difference between both definitions (green curve next to or behind the red one).

In the following, we consider the network generated with base pattern $b_{init} = (11011)$ after four iterative steps ($n = 4$), resulting in a ring network of $N = 5^4 + 1 = 626$ nodes. Our choice is motivated on the one hand by previous studies of chimera states in nonlocally coupled networks [25, 27], where it has been shown that an intermediate coupling range is crucial for the observation of chimera states, too large and too small numbers of connections make this impossible. On the other hand, it has been demonstrated that hierarchical networks with higher clustering coefficient promote chimera states [26, 45]. As we have mentioned before, fractal structures in nature often exhibit a hierarchal level $n = 3$ - 5, therefore, our choice of $n = 4$ is also in this context reasonable. For the fractal topology considered here the clustering coefficient C' in Eq. (5.5) is calculated as $C' = 0.428$ with an effective coupling radius $\bar{r} \approx 0.20$ (or $\rho = \frac{256}{626} \approx 0.41$), which is much smaller than the coupling radius r for which chimeras have been observed in regular nonlocally coupled networks [25, 27]. This is remarkable, since it implies that not only the number of connections is essential for the observation of chimeras, but also their distribution in the network.

Fig. 5.3 Mismatch between clustering coefficients C and C': for different base patterns (colored) and hierarchical steps m the clustering coefficient is calculated using Eq. (5.4) in (**a**) and Eq. (5.5) in (**b**). Panel **a** is reproduced from [45], with the permission of American Institute of Physics (AIP) Publishing

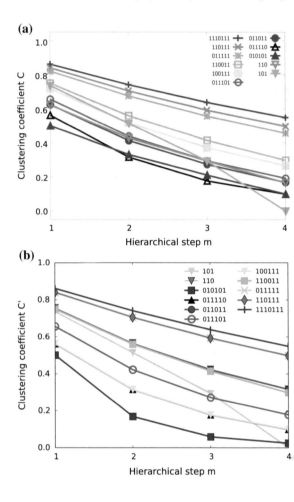

5.2   Influence of Time Delay

In Sect. 2.1.3, we have discussed delay differential equations as an instrument to model propagation or processing time in dynamical systems. Therefore, moving towards more realistic models should include time delays, however, adding time delay drastically increases the dimensionality of a system making the analysis more demanding. In has been shown that time delay generally results in spatial modulation of chimera patterns and, therefore, appearance of clustered chimera states [40]. Since delay differential equations are analogous to space-time systems if the delay interval is interpreted as pseudo-space, delayed feedback systems also exhibit complex self-organized partially coherent, partially incoherent dynamics as an analogy to chimera states [20, 21].

In the last Sect. 5.1, we have introduced a fractal connectivity in a ring network. Let us now turn to the dynamics of each individual oscillator in Eq. (5.1) and study the interplay of time delay and local dynamics in this complex network. We have chosen two paradigmatic models for the local dynamics. Section 5.2.1 closely follows [35], whereas Sects. 5.2.2 and 5.2.3 are related to [36].

5.2.1 Network of Van der Pol Oscillators

We have introduced the Van der Pol oscillator already in Chap. 3 modeling an acoustic oscillator. Originally used to describe limit cycles in electrical circuits, this model has a wide range of applicability. For this so-called relaxation-type oscillator the dynamics of each individual oscillator in Eq. (5.1) is governed by

$$\mathbf{F}(\mathbf{x}) = \begin{pmatrix} v \\ \varepsilon(1 - u^2)v - u \end{pmatrix}, \tag{5.8}$$

where ε denotes the bifurcation parameter. The uncoupled Van der Pol oscillator has a stable fixed point at $\mathbf{x}^* = 0$ for $\varepsilon < 0$ and undergoes a Hopf bifurcation at $\varepsilon = 0$. In this Section, only $\varepsilon = 0.1$ is considered. The interaction in Eq. (5.1) is realized through diffusive coupling with coupling matrix

$$\mathbf{H} = \begin{pmatrix} 0 & 0 \\ b_1 & b_2 \end{pmatrix} \tag{5.9}$$

and real interaction parameters b_1 and b_2. In accordance with Omelchenko et al. [27], throughout the current Chapter we fix the parameters $b_1 = 1.0$ and $b_2 = 0.1$ (see Sect. 4.1.2 for more details about the relevance of the phase-lag parameter).

Figure 5.4 demonstrates chimera states in the system of Eqs. (5.1) and (5.8) for $b_{init} = (11011)$, $n = 4$, $N = 626$, $\varepsilon = 0.1$, and $\sigma = 0.35$, without time delay $\tau = 0$, obtained numerically for symmetric chimera-like initial conditions. We analyze space-time plot (upper panel), the final snapshot of variables u_k at $t = 1000$ (middle panel), and frequencies of oscillators averaged over time window $\Delta T = 10,000$ (bottom panel). Oscillators from coherent domains are phase-locked and have equal mean frequencies. Arc-like profiles of mean frequencies for oscillators from incoherent domain are typical for chimera states.

To uncover the influence of time delay introduced in the coupling term in system of Eqs. (5.1) and (5.8), we analyze numerically the parameter plane of coupling strength σ and delay time τ. Fixing network parameters $b_{init} = (11011)$, $n = 4$, $N = 626$, and $\varepsilon = 0.1$, we choose the chimera pattern of the undelayed system (shown in Fig. 5.4) as initial condition, and vary the values of σ and τ. In numerical simulations of chimera states, the choice of initial conditions often plays a very

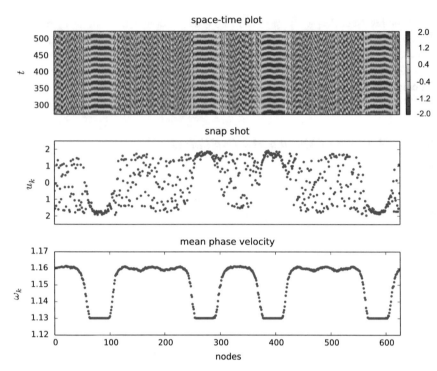

Fig. 5.4 Chimera state in the undelayed case $\tau = 0$ for $b_{init} = (11011)$, $n = 4$, $N = 626$, $\varepsilon = 0.1$, and $\sigma = 0.35$. Note the nonidentical sizes of incoherent domains. The three panels correspond to the same simulation: space-time plot of u (upper panels), snapshots of variables u_k at $t = 1000$ (middle panels), and mean phase velocity profile ω_k (bottom panels). This asymmetric pattern is used as initial condition for further simulations with $\tau \neq 0$. Figure from [35]

important role. Usually, chimera states coexist with the fully synchronized state or coherent traveling waves, and random initial conditions rarely result in chimera patterns. In contrast, specially prepared initial conditions which combine coherent and incoherent spatial domains, increase the probability of observing chimeras. Nevertheless, it is remarkable that the asymmetric structure in Fig. 5.4 evolves from symmetric initial conditions.

Figure 5.5 demonstrates the map of regimes in the parameter plane (τ, σ). In the undelayed case $\tau = 0$ we observe the chimera state shown in Fig. 5.4. The introduction of small time delay for weak coupling strength ($\sigma < 0.3$) immediately destroys the chimera pattern and the incoherent domains characterized by chaotic dynamics appear (yellow dotted region). Nevertheless, for larger values of coupling strength ($\sigma > 0.3$) chimera states are still present. With increasing delay τ we observe a sequence of tongue-like regions (shown red) for chimera states. These regions appear in between large areas of alternating coherent structures: fully synchronized states (yellow regions with horizontal stripes) and traveling waves (yellow regions with diagonal stripes). Closer inspection of the chimera tongues shows that

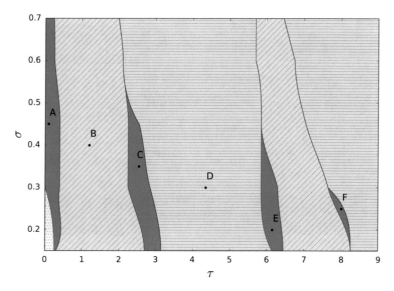

Fig. 5.5 Dynamical scenarios: Chimera tongues (red), in-phase synchronization (horizontally striped yellow region) and coherent traveling waves (diagonally striped yellow region) in the parameter plane (τ, σ) for $b_{init} = (11011)$, $n = 4$, $N = 626$, $\varepsilon = 0.1$. At the transition to a chimera region we can observe chaos (dotted gray region at small τ, σ). The coherent regions are detected by analyzing the mean phase velocity and a snapshot of variables u_k: In the case of equal mean phase velocities and variables u_k we obtain in-phase synchronization, for equal mean phase velocities but unequal variables u_k we obtain coherent traveling waves. The boundary of these regions are fitted by Bézier curves after the (σ, τ) plane is sampled in steps $\Delta\sigma = 0.05$ and $\Delta\tau = 0.05$ - 0.15. Figure from [35]

increasing τ reduces the size of the tongues, and also decreases the maximal σ values, for which chimera states are observed. Moreover, one can easily see that chimera regions appear at τ values close to integer multiples of π.

The sequence of tongues for chimera states in the (τ, σ) parameter plane of Eqs. (5.1) and (5.8) shown in Fig. 5.5 can be understood as a resonance effect in τ. The intrinsic angular frequency of the uncoupled system for small ε is $\omega = 1$ which corresponds to a period of 2π. In many delay systems one expects resonance effects if the delay is an integer or half-integer multiple of this period [10, 48]. The undelayed part of the coupling term in Eqs. (5.1) and (5.8) can be rewritten as $-\sigma u$, neglecting $b_2 = 0.1 \ll b_1 = 1$. This amounts to a rescaling of the uncoupled angular frequency $\omega = 1$ to $\sqrt{1 + \sigma}$ for small ε in the limit of the harmonic oscillator equation $\ddot{u} + (1 + \sigma)u = 0$. Thus the intrinsic period of the coupled system can be roughly approximated by $2\pi/\sqrt{1 + \sigma}$. In the case of point F ($\sigma = 0.25$) in Fig. 5.5, we expect the third chimera tongue around $\tau = 3\pi/\sqrt{1 + 0.25} \approx 8.4$ which agrees reasonably well with the numerical result (see Fig. 5.6 lower panels). Therefore, chimera tongues are shifted to the left for increasing coupling strength σ.

Let us take a closer look at the dynamics inside the tongues. For the parameter values chosen inside the first, leftmost and largest, tongue we find chimera states

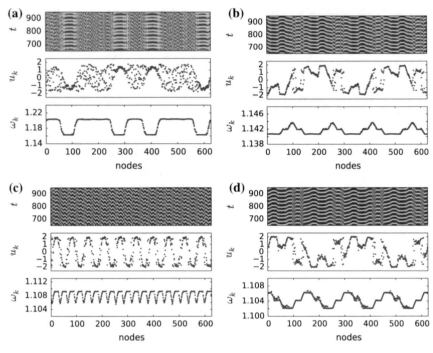

Fig. 5.6 Patterns taken from the chimera tongues in Fig. 5.5 with $b_{init} = (11011)$, $n = 4$, $N = 626$, $\varepsilon = 0.1$: Space-time plot of u (upper panels), snapshots of variables u_k (middle panels), and mean phase velocity profile ω_k (bottom panels) for **a** $\tau = 0.1$ and $\sigma = 0.45$ (point A), **b** $\tau = 2.55$ and $\sigma = 0.35$ (point C), **c** $\tau = 6.15$ and $\sigma = 0.20$ (point E), and **d** $\tau = 8.1$ and $\sigma = 0.25$ (point F). Figure from [35]

similar to the initial condition in Fig. 5.4. In the second and the forth tongue nested chimera structures can be observed (see Fig. 5.6b, d). In the third tongue for $\tau \approx 2\pi$ multichimera states can be observed, e.g., a 20-chimera in Fig. 5.6c. Therefore, the appropriate choice of time delay τ in the system allows one to achieve the desired chimera pattern.

In the parameter plane of time delay and coupling strength the region corresponding to coherent states is dominating (yellow regions in Fig. 5.5). On the one hand we observe the in-phase synchronization regime (see Fig. 5.7b) which is enlarged for increasing coupling strength. On the other hand, we also detect a region of coherent traveling waves with wavenumber $k_w > 1$ (see Fig. 5.7a). Varying the delay time τ allows not only for switching between these states, but also for controlling the speed of traveling waves: in the diagonal striped region in Fig. 5.5 the mean phase velocity decreases for increasing time delays. The pyramidal or conical structure of the mean phase velocity profile in Figs. 5.6b and d is due to the fact that the whole chimera structure is traveling. The speed of traveling is sensitive to the coupling strength and time delay. For a pronounced profile of the mean phase velocity this speed must be small. Otherwise the profile is smeared out over time. An exemplary phase-time

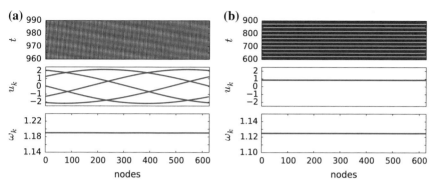

Fig. 5.7 Patterns taken from the coherent (yellow) regions in Fig. 5.5 with $b_{init} = (11011)$, $n = 4$, $N = 626$, $\varepsilon = 0.1$: Space-time plot of u (upper panels), snapshots of variables u_k (middle panels), and mean phase velocity profile ω_k (bottom panels) for **a** $\tau = 1.20$ and $\sigma = 0.4$ (point B), and **b** $\tau = 4.35$ and $\sigma = 0.3$ (point D). Figure from [35]

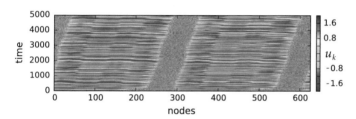

Fig. 5.8 Traveling chimeras: space-time plot of variable u_k in Eqs. (5.1) and (5.8) for the scenario of a traveling chimera structure for $\varepsilon = 0.1$, $\sigma = 0.41$. The initial condition ($t = 0$) are chosen as a 2-chimera state. Other parameters are given by $b_{init} = (11011)$, $n = 4$, $N = 626$, $\tau = 0$

plot of such traveling chimera structure is shown in Fig. 5.8: As we can see in the space-time plot of variable u, the chimera state is traveling clockwise on the ring. For $2000 > t > 3000$ this speed is almost constant, whereas for the time interval $2000 < t < 3000$ the speed decreases rapidly and the position of the chimera state remains at a fixed position on the ring.

5.2.2 A Model for Neuronal Networks

The Van der Pol oscillator applied in the last Sect. 5.2.1 has been extended by Fitzhugh and Nagumo as a model in biological sciences describing the action potentials of neurons. Since then the FitzHugh-Nagumo oscillator is a paradigmatic model for neural systems [9, 23], but is also used to describe chemical [42] and optoelectronic [34] oscillators and nonlinear electronic circuits [11].

The local dynamics for the variable $\mathbf{x}_k = (u_k, v_k)^T \in \mathbb{R}^2$ in Eq. (5.1), where u_k and v_k are the activator and inhibitor variables, respectively, are given by

$$\mathbf{F(x)} = \begin{pmatrix} \varepsilon^{-1}(u - \frac{u^3}{3} - v) \\ u + a \end{pmatrix}, \tag{5.10}$$

where $\varepsilon > 0$ is a small parameter characterizing a time scale separation, which we fix at $\varepsilon = 0.05$ throughout the current Chapter. Depending on the threshold parameter a, the FitzHugh-Nagumo oscillator exhibits either oscillatory ($|a| < 1$) or excitable ($|a| > 1$) behavior. We consider the oscillatory regime ($a = 0.5$) in this Chapter. In contrast to Eq. (5.9), the interaction in Eq. (5.1) is realized here through diffusive coupling with coupling matrix

$$\mathbf{H} = \begin{pmatrix} \varepsilon^{-1}\cos\phi & \varepsilon^{-1}\sin\phi \\ -\sin\phi & \cos\phi \end{pmatrix}. \tag{5.11}$$

In accordance with Sect. 4.1.2, throughout the current Chapter we fix the coupling phase $\phi = \frac{\pi}{2} - 0.1$.

Figure 5.9 demonstrates a chimera state (20-chimera) in the system of Eqs. (5.1) and (5.10) for $b_{init} = (11011)$, $n = 4$, $N = 626$, $\varepsilon = 0.05$, $\phi = \frac{\pi}{2} - 0.1$, $\sigma = 0.05$, and time delay $\tau = 3.6$, obtained numerically for initial conditions, randomly distributed on the circle $u_k^2 + v_k^2 = 4$. We call such a distribution random initial conditions. We analyze the space-time plot (upper panel), the final snapshot of the activator variables u_k at $t = 50,000$ (middle panel), and the phase velocities ω_k of

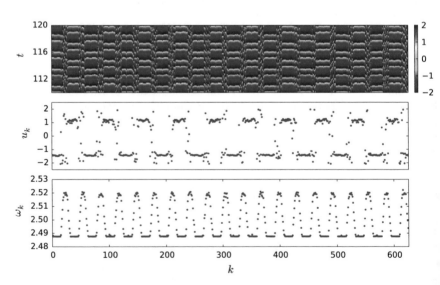

Fig. 5.9 Chimera state in the case $\tau = 3.6$ and $\sigma = 0.05$ for $b_{init} = (11011)$, $n = 4$, $N = 626$, $a = 0.5$, $\varepsilon = 0.05$, and $\phi = \frac{\pi}{2} - 0.1$. Random initial conditions are used. The three panels correspond to the same simulation: space-time plot of u_k (upper panel), snapshot of variable u_k at $t = 50,000$ (middle panel), and mean phase velocity profile ω_k (bottom panel). Figure from [36]

the oscillators (bottom panel). These mean phase velocities of the oscillators are calculated as $\omega_k = 2\pi S_k / \Delta T$, $k = 1, ..., N$, where S_k denotes the number of complete rotations realized by the kth oscillator during the time ΔT. Throughout the current Chapter we used $\Delta T = 10{,}000$. Oscillators from coherent domains are phase-locked and have equal mean frequencies. Arc-like profiles of the mean phase velocities for oscillators from the incoherent domain are typical for chimera states.

To uncover the influence of time delay introduced in the coupling term in Eqs. (5.1) and (5.10), we analyze numerically the parameter plane of coupling strength σ and delay time τ. Fixing the network parameters $b_{init} = (11011), n = 4, N = 626, a = 0.5$, and $\varepsilon = 0.05$, we choose random initial conditions, and vary the values of σ and τ.

Figure 5.10 shows the map of regimes in the parameter plane (τ, σ). In the undelayed case $\tau = 0$ we cannot observe chimera states for random initial conditions. The introduction of small time delay for weak coupling strength does not change the behavior and the system stays in the completely incoherent regime characterized by chaotic dynamics (grey dotted region). Nevertheless, for larger values of coupling strength σ chimera states can be observed for small τ. With increasing delay τ we observe a sequence of tongue-like regions, which are bounded by red curves, on

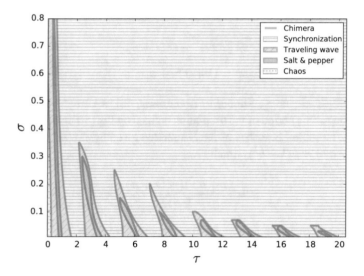

Fig. 5.10 Chimera tongues: Chimeras occur on the boundaries (red curves) between in-phase synchronization (horizontally striped blue region), coherent traveling waves (diagonally striped yellow-green regions), and "salt and pepper" dynamics (dotted red regions) in the parameter plane (τ, σ). Below the first traveling wave region we can observe chaos (dotted grey region at small τ, σ). Random initial conditions are used for all numerical simulations. Other parameters as in Fig. 5.9. Figure modified from [36]

which chimera states occur. These regions appear in between larger areas of coherent structures: fully synchronized states (blue regions with horizontal stripes) alternating with coherent traveling waves (yellow-green regions with diagonal stripes). Inside the tongues we can observe "salt and pepper" states, which are characterized by strong variations on very short length scales [1, 39] (red dotted regions). Closer inspection of the chimera tongues shows that increasing τ reduces the size of the tongues, and also decreases the maximum values of σ for which chimera states are observed. Moreover, one can easily see that chimera regions appear at τ values close to half-integer multiples of the period of the uncoupled system $T \approx 2.3$.

5.2.3 Analytical Approach

In many delay systems one expects resonance effects if the delay is an integer or half-integer multiple of the period of the uncoupled system [10, 48]. The undelayed part of the coupling term in Eqs. (5.1) and (5.10) is the most important part in case of incoherence (see tongues in Fig. 5.10) and can be rewritten as follows, neglecting $\cos \phi \ll 1$ and setting $\sin \phi \approx 1$ (it is possible to keep ϕ, but it complicates the algebra, see Eq. (5.18)):

$$\varepsilon \dot{u} = u - \frac{u^3}{3} - (1 + \sigma)v,$$
$$\dot{v} = (1 + \sigma)u + a, \tag{5.12}$$

where $u \equiv u_k$ and $v \equiv v_k$. Similar to Brandstetter [4], we employ an analytic approximation for the period of the oscillation defined by Eq. (5.12). We consider slow motion on the falling branches of the u-nullcline given by $(1 + \sigma)v = u - \frac{u^3}{3}$ and hence $(1 + \sigma)\dot{v} = \dot{u}(1 - u^2)$, which gives:

$$\dot{u} = \frac{(1 + \sigma)^2 u + (1 + \sigma)a}{1 - u^2}. \tag{5.13}$$

It is possible to integrate this equation analytically from $\pm u_+$ to $\pm u_-$, which are approximately the limits of the slow parts of the u-nullcline (see Fig. 5.11a), given by $u_+ = 2$ and $u_- = 1$. With this we obtain a rough approximation of the intrinsic period $T(\sigma)$ of the coupled system, neglecting the fast parts of the trajectory $u(t)$:

$$T(\sigma) \propto (1 + \sigma)^{-2} \left[u_+^2 - u_-^2 + \left(1 - \left(\frac{a}{1 + \sigma} \right)^2 \right) \ln \frac{a^2 - (1 + \sigma)^2 u_-^2}{a^2 - (1 + \sigma)^2 u_+^2} \right]. \tag{5.14}$$

As we can see in Fig. 5.11b, the period T decreases with increasing σ. Therefore, due to the resonance condition of τ with respect to the intrinsic period T, the chimera tongues are shifted to the left with increasing coupling strength σ.

In the case of complete synchronization (blue region in Fig. 5.10) we cannot neglect the delayed terms $v_\tau \equiv v(t - \tau)$ and $u_\tau \equiv u(t - \tau)$ in Eqs. (5.1) and (5.10):

$$\varepsilon \dot{u} = u - \frac{u^3}{3} - v + \sigma (v_\tau - v),$$
$$\dot{v} = u + a - \sigma (u_\tau - u). \tag{5.15}$$

Due to the almost linear behavior on the slow branches (exemplarily shown by the straight connection between $u(t)$ and $u(t - \tau)$ in Fig. 5.11a), we assume $\mathbf{x}(t) - \mathbf{x}(t - \tau) = \tau \dot{\mathbf{x}}(t)$ for values of τ close to multiples of the period mT with $m \in \mathbb{N}$:

$$\varepsilon \dot{u} = u - \frac{u^3}{3} - v - \sigma \tau \dot{v},$$
$$\dot{v} = u + a + \sigma \tau \dot{u}. \tag{5.16}$$

We can insert the second equation into the first one and analyze the dynamics on the falling branches of the u-nullcline given by $v = u - \frac{u^3}{3} - \sigma \tau (u + a)$:

$$\dot{u} = \frac{u + a}{1 - u^2 - 2\sigma \tau}. \tag{5.17}$$

This is an approximation of the equation which would have been obtained if the phase-lag term $\cos \phi$ were not been neglected:

$$\dot{u} = \frac{\frac{u+a}{1+\sigma \tau \cos \phi}}{1 - u^2 - 2\sigma \tau \frac{\sin \phi}{1+\sigma \tau \cos \phi}}. \tag{5.18}$$

In the case of values of τ close to T we can calculate the period of the synchronized oscillations as

$$T_{\text{sync}}(\tau) \propto u_+^2 - u_-^2 + \left(1 - a^2 - 2\sigma \tau\right) \ln \frac{a^2 - u_-^2}{a^2 - u_+^2}. \tag{5.19}$$

As proportionality factor in Eqs. (5.14) and (5.19) we assume $1 + e(\varepsilon)$, where $e(0.05) = 0.3$ is a constant parameter, determined by fitting the analytical solution of Eq. (5.14) for $\sigma = 0$ to the numerical simulation of Eqs. (5.1) and (5.10) for $\sigma = 0$. As shown in [49], delay systems generically have branches of periodic solutions, which are reappearing for integer multiples of the intrinsic period T of the system. A solution for $\tau = \tau_0 < T$ reappears for all values

$$\tau_m = \tau_0 + mT_{\text{sync}}(\tau_0) \tag{5.20}$$

with $m \in \mathbb{N}$, and T_{sync} depends upon τ_0 according to Eq. (5.19). The branches T_{sync} of the synchronized solutions are piecewise linear functions of τ, as shown in

Fig. 5.11c, where $m = 1, 2, \ldots$ numbers the branches. With increasing m the branches are stretched by $\frac{\partial \tau_m}{\partial \tau_0}$ and their slope decreases (see [49]). To take into account this mapping for $m > 0$, τ in Eq. (5.19) has to be replaced by

$$\tau' = \tau_0 \left(\frac{\partial \tau_m}{\partial \tau_0} \right)^{-1} = \tau_0 \left(1 - 2\sigma m (1 + e) \ln \frac{a^2 - u_-^2}{a^2 - u_+^2} \right)^{-1}, \qquad (5.21)$$

where for a given $\tau = \tau_m > T$, τ_0 and m can be calculated from Eq. (5.20). Equation (5.19) now reads

$$T_{\text{sync}}(\tau') = (1 + e) \left[u_+^2 - u_-^2 + \left(1 - a^2 - 2\sigma \tau' \right) \ln \frac{a^2 - u_-^2}{a^2 - u_+^2} \right]. \qquad (5.22)$$

A comparison of this analytical result for the period T_{sync} in the synchronized regime with numerical simulations is given in Fig. 5.11c. Depending on the initial conditions, we can find chimera states in the red shaded regions at the boundaries of the piecewise linear branches, which occur if the delay times τ are half-integer multiples of the intrinsic period $T_{\text{sync}}(0) = T$. They are marked in Fig. 5.11c (red shaded) for $\sigma = 0.15$. Note that the period is piecewise linear as a function of τ and also of σ (in case of $m = 0$ in Eq. (5.21)) in the synchronized regime, whereas it is nonlinear in the non-synchronized regime.

In addition, we can see a decrease of the maximum of the chimera tongues with increasing τ in Fig. 5.10. In [14–16] a similar effect has been observed.

Let us now take a closer look at the dynamics inside the tongues in Fig. 5.10. For the parameter values chosen inside the first, leftmost and largest, tongue we find multichimera states (20-chimera) similar to Fig. 5.9 (see Fig. 5.12a) and nested chimera structures (see Fig. 5.12b). These structures are traveling in space, so that the mean phase velocity profile in the upper panel shows a pyramidal or conical structure. The speed of traveling is sensitive to the coupling strength and time delay. For a pronounced profile of the mean phase velocity this speed should be small. Otherwise it is smeared out over time. Figure 5.12c and d show two examples of the transition region from complete synchronization to chimera states. Also here we have coherent and incoherent domains. In contrast to the other examples, we can find a complex structure of the mean phase velocity profiles (see bottom panels). In general, the appropriate choice of time delay τ in the system allows one to achieve the desired chimera pattern.

In the parameter plane of delay time τ and coupling strength σ the region corresponding to coherent states is dominating (blue and yellow regions in Fig. 5.10). On the one hand, we observe the in-phase synchronization regime (see Fig. 5.13b) which is enlarged for increasing coupling strength. On the other hand, we also detect a region of coherent traveling waves with wavenumber $k_w > 1$ (see Fig. 5.13a) and $k_w < 1$ (see Fig. 5.13c). Varying the delay time τ allows not only for switching between these states, but also for controlling the speed of traveling waves: in the diagonal striped yellow region in Fig. 5.10 the mean phase velocity decreases for increasing time de-

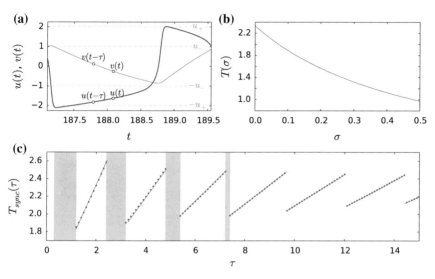

Fig. 5.11 Analytical approximation of the period T for an FitzHugh-Nagumo system with delayed coupling: **a** limit cycle of the variables $u(t)$ (dark blue line) and $v(t)$ (light green line) of a single FitzHugh-Nagumo oscillator with delayed feedback representing the synchronized state of Eqs. (5.1) and (5.10) for $\tau = 2.1$ and $\sigma = 0.15$. The dashed grey lines indicate $\pm u_{\pm}$ respectively, given by $u_{+} = 2$ and $u_{-} = 1$. **b** Period T versus σ of the FitzHugh-Nagumo oscillator given by Eq. (5.14), valid for parameters from the incoherent regimes in Fig. 5.10. As the proportionality factor we assume $1 + e(\varepsilon)$, with $e(0.05) = 0.3$. **c** Period of the synchronized solution T_{sync} versus delay time τ. Comparison of numerics (dots) and analytics (lines), given by Eq. (5.22) for $\sigma = 0.15$. The red shaded regions correspond to the tongues in Fig. 5.10. Other parameters for all panels as in Fig. 5.9. Figure from [36]

lays. In addition, we can observe "salt and pepper" states (see Fig. 5.13d), where all nodes oscillate with the same phase velocity but they are distributed between states with phase-lag π incoherently [1].

5.3 Summary

In the current Chapter, we have analyzed chimera states in ring networks of oscillators with hierarchical connectivities and delayed coupling. In the first Sect. 5.1, we have proceeded from simple nonlocally coupled ring topologies to more complex networks, reflecting the structure of real-world problems. We have introduced hierarchical connectivities and presented the classical Cantor construction algorithm to generate such connectivities. Moreover, we have provided measures as the effective coupling radius and the clustering coefficient to quantify hierarchical connectivities. Inspired by patterns in nature, we have constructed a hierarchical connectivity matrix with a hierarchical level of $n = 4$. In Sect. 5.2, we have provided a numerical

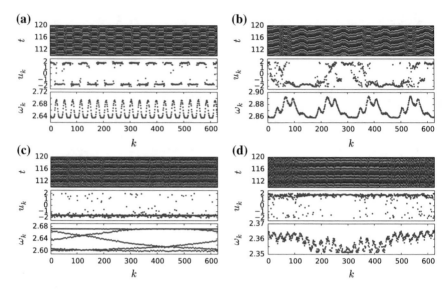

Fig. 5.12 Patterns occurring in the chimera tongues in Fig. 5.10: space-time plot of u (upper panels), snapshot of variables u_k (middle panels), and mean phase velocity profile ω_k (bottom panels) for **a** $\tau = 1.0$ and $\sigma = 0.1$, **b** $\tau = 1.0$ and $\sigma = 0.15$, **c** $\tau = 1.4$ and $\sigma = 0.05$, and **d** $\tau = 10.7$ and $\sigma = 0.01$. Other parameters as in Fig. 5.9. Figure from [36]

study of complex spatio-temporal patterns in the network for a fixed base pattern. The effective coupling radius for this pattern is given by $\bar{r} \approx 0.20$ and would not allow for chimera states in case of a non-local coupling topology. Furthermore, our study has been focused on the role of time delay in the coupling term and its influence on the chimera states. To prove the generality of our observations, we have used two different paradigmatic models for the oscillators: On the one hand, the Van der Pol oscillator in Sect. 5.2.1, on the other hand, the FitzHugh-Nagumo model in Sect. 5.2.2.

In the parameter plane of time delay τ and coupling strength σ, we have determined the stability regimes for different types of chimera states, alternating with regimes of coherent states. An appropriate choice of time delay allows us to stabilize several types of chimera states. The interplay of complex hierarchical network topology and time delay results in a plethora of patterns going beyond regular two-population or nonlocally coupled ring networks: we observe chimera states with coherent and incoherent domains of non-identical sizes and non-equidistantly distributed in space. Moreover, traveling and non-traveling chimera states can be obtained for a proper choice of time delay. We also demonstrate that time delay can induce patterns which are not observed in the undelayed case. In addition, we have shown analytically the influence of τ upon the period; i.e., the phase velocity, a piecewise linear dependence in regimes with coherent states, whereas a nonlinear dependence upon τ is found for incoherent states.

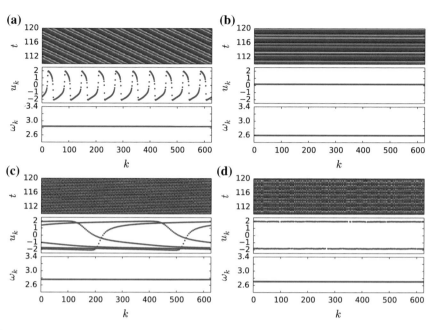

Fig. 5.13 Patterns occurring in non-chimera regimes of Fig. 5.10: space-time plot of u (upper panels), snapshot of variables u_k (middle panels), and mean phase velocity profile ω_k (bottom panels) for **a** $\tau = 1.1$ and $\sigma = 0.15$, **b** $\tau = 4.5$ and $\sigma = 0.1$, **c** $\tau = 5.1$ and $\sigma = 0.1$, and **d** $\tau = 5.5$ and $\sigma = 0.1$. Other parameters as in Fig. 5.9. Figure from [36]

Our analysis has shown that networks with complex hierarchical topologies, as arising in neuroscience, can exhibit diverse nontrivial patterns. Time delay can play the role of a powerful control mechanism which allows either to promote or to destroy chimera patterns in a single-layer system. It has been recently studied how chimera states resemble some phenomena in the nervous system such as unihemispheric sleep in birds [33] or epilepsy [13]. Passing such patterns from a single layer to another therefore might also act an important role in the nervous system and will be the focus of the next Chapters.

References

1. Bachmair CA, Schöll E (2014) Nonlocal control of pulse propagation in excitable media. Eur Phys J B **87**:276
2. Bick C, Martens EA (2015) Controlling chimeras. New J Phys **17**:033030
3. Böhm F, Zakharova A, Schöll E, Lüdge K (2015) *Amplitude-phase coupling drives chimera states in globally coupled laser networks*. Phys Rev E **91**:040901(R)
4. Brandstetter SA, Dahlem MA, Schöll E (2010) Interplay of time-delayed feedback control and temporally correlated noise in excitable systems. Philos Trans R Soc A **368**:391

5. Chouzouris T, Omelchenko I, Zakharova A, Hlinka J, Jiruska P, Schöll E (2018) Chimera states in brain networks: empirical neural vs. modular fractal connectivity. Chaos **28**:045112

6. Crook SM, Ermentrout GB, Vanier MC, Bower JM (1997) The role of axonal delay in the synchronization of networks of coupled cortical oscillators. J Comput Neurosci **4**:161–172

7. Expert P, Evans TS, Blondel VD, Lambiotte R (2011) Uncovering space-independent communities in spatial networks. Proc Natl Acad Sci USA **108**:7663

8. Feder J (1988) Fractals. Plenum Press, New York

9. FitzHugh R (1961) Impulses and physiological states in theoretical models of nerve membrane. Biophys J **1**:445–466

10. Hövel P, Schöll E (2005) Control of unstable steady states by time-delayed feedback methods. Phys Rev E **72**:046203

11. Heinrich M, Dahms T, Flunkert V, Teitsworth SW, Schöll E (2010) Symmetry breaking transitions in networks of nonlinear circuit elements. New J Phys **12**:113030

12. Hizanidis J, Panagakou E, Omelchenko I, Schöll E, Hövel P, Provata A (2015) Chimera states in population dynamics: networks with fragmented and hierarchical connectivities. Phys Rev E **92**:012915

13. Jiruska P, de Curtis M, Jefferys JGR, Schevon CA, Schiff SJ, Schindler K (2013) Synchronization and desynchronization in epilepsy: controversies and hypotheses. J Physiol **591**(4):787–797

14. Just W, Bernard T, Ostheimer M, Reibold E, Benner H (1997) Mechanism of time-delayed feedback control. Phys Rev Lett **78**:203

15. Just W, Reibold E, Kacperski K, Fronczak P, Holyst JA, Benner H (2000) Influence of stable Floquet exponents on time-delayed feedback control. Phys Rev E **61**:5045

16. Just W, Benner H, Cv Loewenich (2004) On global properties of time-delayed feedback control: weakly nonlinear analysis. Phys D **199**:33

17. Katsaloulis P, Verganelakis DA, Provata A (2009) Fractal dimension and lacunarity of tractography images of the human brain. Fractals **17**:181–189

18. Katsaloulis P, Ghosh A, Philippe AC, Provata A, Deriche R (2012) Fractality in the neuron axonal topography of the human brain based on 3-D diffusion MRI. Eur Phys J B **85**:1–7

19. Katsaloulis P, Hizanidis J, Verganelakis DA, Provata A (2012) Complexity measures and noise effects on diffusion magnetic resonance Imaging of the neuron axons network in human brains. Fluctuation Noise Lett **11**:1250032

20. Larger L, Penkovsky B, Maistrenko Y (2013) Virtual chimera states for delayed-feedback systems. Phys Rev Lett **111**:054103

21. Larger L, Penkovsky B, Maistrenko Y (2015) Laser chimeras as a paradigm for multistable patterns in complex systems. Nat Commun **6**:7752

22. Mandelbrot BB (1983) The fractal geometry of nature, 3rd edn. WH Freeman and Comp, New York

23. Nagumo J, Arimoto S, Yoshizawa S (1962) An active pulse transmission line simulating nerve axon. Proc IRE **50**:2061–2070

24. Newman MEJ (2003) The structure and function of complex networks. SIAM Rev **45**:167–256

25. Omelchenko I, Omel'chenko OE, Hövel P, Schöll E (2013) When nonlocal coupling between oscillators becomes stronger: patched synchrony or multichimera states. Phys Rev Lett **110**:224101

26. Omelchenko I, Provata A, Hizanidis J, Schöll E, Hövel P (2015) Robustness of chimera states for coupled FitzHugh-Nagumo oscillators. Phys Rev E **91**:022917

27. Omelchenko I, Zakharova A, Hövel P, Siebert J, Schöll E (2015) Nonlinearity of local dynamics promotes multi-chimeras. Chaos **25**:083104

28. Omelchenko I, Omel'chenko OE, Zakharova A, Wolfrum M, Schöll E (2016) Tweezers for chimeras in small networks. Phys Rev Lett **116**:114101

29. Omelchenko I, Omel'chenko OE, Zakharova A, Schöll E (2018) Optimal design of Tweezer control for chimera states. Phys Rev E **97**:012216

30. Omelchenko I, Hülser T, Zakharova A, Schöll E (2019) Control of chimera states in multilayer networks. Front Appl Math Stat **4**:67

31. Panaggio MJ, Abrams DM (2015) Chimera states: coexistence of coherence and incoherence in networks of coupled oscillators. Nonlinearity **28**:R67

32. Provata A, Katsaloulis P, Verganelakis DA (2012) Dynamics of chaotic maps for modelling the multifractal spectrum of human brain diffusion tensor images. Chaos Solitons Fractals **45**:174–180

33. Rattenborg NC, Voirin B, Cruz SM, Tisdale R, Dell'Omo G, Lipp HP, Wikelski M, Vyssotski AL (2016) Evidence that birds sleep in mid-flight. Nat Commun **7**:12468

34. Rosin DP, Callan KE, Gauthier DJ, Schöll E (2011) Pulse-train solutions and excitability in an optoelectronic oscillator. Europhys Lett **96**:34001

35. Sawicki J, Omelchenko I, Zakharova A, Schöll E (2017) Chimera states in complex networks: interplay of fractal topology and delay. Eur Phys J Spec Top **226**:1883–1892

36. Sawicki J, Omelchenko I, Zakharova A, Schöll E (2019) Delay-induced chimeras in neural networks with fractal topology. Eur Phys J B **92**:54

37. Schmidt L, Krischer K (2015) Chimeras in globally coupled oscillatory systems: from ensembles of oscillators to spatially continuous media. Chaos **25**:064401

38. Schmidt L, Krischer K (2015) Clustering as a prerequisite for chimera states in globally coupled systems. Phys Rev Lett **114**:034101

39. Semenova N, Strelkova G, Anishchenko VS, Zakharova A (2017) Temporal intermittency and the lifetime of chimera states in ensembles of nonlocally coupled chaotic oscillators. Chaos **27**:061102

40. Sen A, Dodla R, Johnston G, Sethia GC (2010) Amplitude death, synchrony, and chimera states in delay coupled limit cycle oscillators. In: Atay FM (ed) Complex time-delay systems, vol 16. understanding complex systems. Springer, Berlin, pp 1–43

41. Sethia GC, Sen A (2014) Chimera states: the existence criteria revisited. Phys Rev Lett **112**:144101

42. Shima S, Kuramoto Y (2004) Rotating spiral waves with phase-randomized core in nonlocally coupled oscillators. Phys Rev E **69**:036213

43. Sieber J, Omel'chenko OE, Wolfrum M (2014) Controlling unstable chaos: stabilizing chimera states by feedback. Phys Rev Lett **112**:054102

44. Tsigkri-DeSmedt ND, Hizanidis J, Hövel P, Provata A (2016) Multi-chimera states and transitions in the leaky integrate-and-fire model with excitatory coupling and hierarchical connectivity. Eur Phys J ST **225**:1149

45. Ulonska S, Omelchenko I, Zakharova A, Schöll E (2016) Chimera states in networks of Van der Pol oscillators with hierarchical connectivities. Chaos **26**:094825

46. Watanabe S, Strogatz SH (1993) Integrability of a globally coupled oscillator array. Phys Rev Lett **70**:2391

47. Watts DJ, Strogatz SH (1998) Collective dynamics of 'small-world' networks. Nature **393**:440–442

48. Yanchuk S, Wolfrum M, Hövel P, Schöll E (2006) Control of unstable steady states by long delay feedback. Phys Rev E **74**:026201

49. Yanchuk S, Perlikowski P (2009) Delay and periodicity. Phys Rev E **79**:046221

50. Yeldesbay A, Pikovsky A, Rosenblum M (2014) Chimeralike states in an ensemble of globally coupled oscillators. Phys Rev Lett **112**:144103

51. zur Bonsen A, Omelchenko I, Zakharova A, Schöll E (2018) Chimera states in networks of logistic maps with hierarchical connectivities. Eur Phys J B **91**:65

Part II
Multilayer Systems

Chapter 6
Partial Synchronization in 2-Community Networks

Large interconnected systems built from individual nodes with complex dynamics are common in many seemingly distinct fields of natural sciences, technology, and economy. In the last Chaps. 4 and 5, we have investigated the interplay of dynamics and delay in ring networks. These networks are single-layer networks. Multilayer networks can give a general framework to describe and model real life examples of such systems, as for instance interactions between genes, proteins, and neurons, between transportation systems, and between social networks [14]. As introduced in Sect. 2.2.2, a common property of the mentioned examples are that they can be modeled with a network consisting of separate layers (multilayer network) in which the nodes are connected with different types of links (intra-layer) than those in between the layers (inter-layer). Next to the topology of the mentioned systems, understanding their general spatiotemporal patterns as well as the synchronization and desynchronization (in between separate layers) of these patterns can be crucial in many real life applications [50]. Understanding synchronization patterns in such model networks inside and in between specific layers can help us better understanding problems in nature and technology and it may allow us to create and study computational models for as complex systems as interconnected regions of the mammalian brain.

In this Chapter, we analyze partial synchronization patterns in a network of FitzHugh-Nagumo oscillators with empirical structural connectivity measured in healthy human subjects. We report a dynamical asymmetry between the hemispheres, induced by the natural structural asymmetry. We show that the dynamical asymmetry can be enhanced by introducing the inter-hemispheric coupling strength as a control parameter for partial synchronization patterns. We specify the possible modalities for existence of unihemispheric sleep in human brain, where one hemisphere sleeps while the other remains awake. In fact, this state is common among migratory birds and mammals like aquatic species. This Chapter includes material from [35, 36].

The outline of the Chapter is as follows: In Sect. 6.1, we introduce empirical structural connectivity measured in human subjects and explain applicable measures. In Sect. 6.2, we present appropriate measures for asymmetries and analyze structural

© Springer Nature Switzerland AG 2019
P. Sawicki, *Delay Controlled Partial Synchronization in Complex Networks*,
Springer Theses, https://doi.org/10.1007/978-3-030-34076-6_6

and dynamical asymmetries in the hemispheric network. In Sect. 6.3, we describe partial synchronization scenarios and apply these patterns as a model for unihemispheric sleep. We conclude with Sect. 7.6 summarizing our results.

6.1   Modeling Brain Dynamics

A well-known phenomenon in nature is unihemispheric slow-wave sleep, exhibited by aquatic mammals including whales, dolphins and seals, and multiple bird species. Unihemispheric sleep, as the name suggests, is the remarkable ability to engage in deep (slow-wave) sleep with a single hemisphere of the brain while the other hemisphere remains awake [25, 38, 39]. Interestingly, sleep and wakefulness are characterized by a high and low degree of synchronization, respectively [38]. Sleep is associated with specific synchronized oscillations, i.e., sleep spindles and slow oscillations in the thalamocortical system [52]. In addition, arousal- and sleep-promoting neural assemblies undergo collective activity resulting in secretion of sleep-regulating neurotransmitters [46]. While the synchronization processes can differ between adults and children [51], transitions from wakefulness to sleep are widely accompanied by synchronization phenomena [28]. In the human brain the first-night effect, which describes troubled sleep in a novel environment, has been related to asymmetric dynamics recently, i.e., a manifestation of one hemisphere of the brain being more vigilant than the other [54].

Sleep is a dynamical macrostate of the brain that is observed over a wide range of animal species. Sleep is accompanied by a loss of consciousness and conscious perceptions, and muscle activity is reduced or absent. As we will explain in Sect. 6.1.2 more accurately, sleep alternates between rapid-eye-movement (REM) and non-REM stages N1, N2, N3, the latter dominated by slow oscillations (1 Hz and below) which can also emerge locally [23, 58]. Sleep stage switching dynamics includes wake/sleep asymmetric stochasticity [44], but obeys an underlying control by regulatory circuits forming bistable biological flipflop switches [7, 16, 32, 42, 43], and sleep regulation is coupled to the sleep oscillations of the thalamocortical system [10]. While most animals follow a similar qualitative sleep pattern and fall into sleep with both hemispheres, in certain bird and mammal species, seals and dolphins sleep can be unihemispheric [25].

As different species are assumed to share a similar regulatory circuitry [25], one may hypothesize that both unihemispheric and bihemispheric sleep are possible dynamical states of the same network, where the symmetry breaking may be induced by small anatomical asymmetries, the first night effect [4, 54] or by additional mechanisms that evolved under pressure of selection [37, 40].

It has been speculated that unihemispheric sleep is related to the spontaneous symmetry-breaking phenomenon of chimera states in oscillator networks [3, 29]; those states combine spatially coexisting domains of synchronized and desynchronized dynamics [2, 22, 33, 47, 49].

While the neurophysiological processes that ensure the existence of this dynamical state of unihemispheric sleep remain largely unknown, it is presumed that a certain degree of structural interhemispheric separation is a necessary condition for this pattern to persist. Therefore, we propose to model unihemispheric sleep by a two-community network of the two hemispheres where the inter-hemispheric coupling strength is smaller than the intra-hemispheric coupling. In this Section, we introduce empirical structural brain networks which can be obtained from diffusion-weighted magnetic resonance imaging. We model the spiking dynamics of the neurons by the paradigmatic FitzHugh-Nagumo model, and investigate possible partial synchronization patterns. Furthermore, we explain how unihemispheric slow-wave sleep, a particular partial synchronization pattern in the brain, can be measured by means of electroencephalography (EEG).

Electroencephalography is a medical diagnosis method to represent electrical activity in the brain. By placing electrodes on the scalp, this measurement displays the electrical activity of the brain by summing up voltage fluctuations at head surface. The reason for these fluctuations are physiological processes of neurons in a brain region, which the corresponding electrode is attached to. Therefore, the signals of electroencephalography can be interpreted as the sum over a given brain region [11, 34]. The measurements given by an electroencephalography has a quite wide range of applications and is used, inter alia, to distinguish epileptic seizures, to evaluate head injury, and to track brain changes during different sleep phases (see Sect. 6.1.2).

6.1.1 Empirical Structural Brain Networks

We consider an empirical structural brain network shown in Fig. 6.1 where every region of interest is modeled by a single FitzHugh-Nagumo oscillator. The brain network has been obtained from diffusion-weighted magnetic resonance imaging data measured in healthy human subjects as part of a larger study focusing on connectivity changes in schizophrenia. For details of the measurement procedure including acquisition parameters, see [27], for previous utilization of the structural networks to analyze chimera states see [9]. The data have been analyzed by means of probabilistic tractography [5] as implemented in the FMRIB Software Library, where FMRIB stands for Functional Magnetic Resonance Imaging of the Brain (for further details see www.fmrib.ox.ac.uk/fsl/). The anatomic network of the cortex and subcortex is measured using Diffusion Tensor Imaging (DTI) and afterwards divided into 90 predefined regions according to the Automated Anatomical Labeling (AAL) atlas [57]. Each node of the network corresponds to a specific brain region. Indirect information of the white matter fibers connecting different brain regions is provided by diffusion-weighted Magnetic Resonance Imaging (dMRI) measuring the preferred diffusion direction in each voxel of the brain. Probabilistic tractography [5] subsequently provides for each voxel a set of $n_s = 5000$ streamlines, simulating the possible white matter fiber tracts. A coefficient P_{kj} gives the connectivity probability from the kth to the jth region and is calculated by the proportion of streamlines connecting voxels in region k to voxels of region j on the condition that they originate in region k. Thus, a weighted adjacency matrix of size 90×90, with node indices $k \in N = \{1, 2, ..., 90\}$ is constructed. Finally, each entry in this adjacency matrix, i.e., the connectivity between every two regions is averaged over 20 subjects (mean age 33 years, standard deviation 5.7 years, 10 females, 2 left-handed) yielding the average empirical structural brain network $\mathbf{A} = \{A_{kj}\}$. For more detailed information and further references we refer to [9].

The pipeline for constructing the structural network has been adopted from previous studies of differences in connectivity patterns between healthy subjects and schizophrenia patients [8]. Obtaining such connectivity information using diffusion tractography is known to face a range of challenges [17, 45]. While some estimates of the strength and direction of structural connections from measurements of brain activity can in principle be attempted, the relation of these can vary with parameters of the local dynamics and coupling function [17]. Note that in contrast to the original Automated Anatomical Labeling (AAL) indexing, where sequential indices correspond to homologous brain regions, the indices in Fig. 6.1 are rearranged such that $k \in N_L = \{1, 2, ..., 45\}$ corresponds to left and $k \in N_R = \{46, ..., 90\}$ to the right hemisphere. Thereby the hemispheric structure of the brain, i.e., stronger intra-hemispheric coupling compared to inter-hemispheric coupling, is highlighted (see Fig. 6.1a). Furthermore, note that there is a very slight structural asymmetry of the two brain hemispheres. This asymmetry holds for each of the 20 considered brain networks [35]. In Fig. 6.1a, we can notice the difference for example by comparing the matrix elements at $k = j \approx 20$ with their counterparts in the second hemisphere at $k = j \approx 65$. This structural asymmetry is well described in the neuroscience lit-

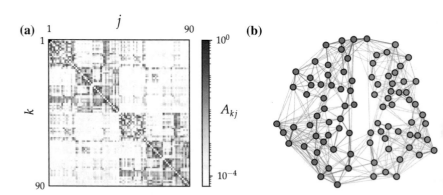

Fig. 6.1 Model for the hemispheric brain structure: **a** weighted adjacency matrix $\mathbf{A} = \{A_{kj}\}$ of the averaged empirical structural brain network derived from twenty healthy human subjects by averaging over the coupling between two brain regions k and j. The brain regions k, j are taken from the Automated Anatomic Labeling (AAL) atlas [57], but re-labeled such that $k = 1, ..., 45$ and $k = 46, ..., 90$ correspond to the left and right hemisphere, respectively. **b** Schematic representation of the graph of the brain structure with highlighted left (dark blue) and right (light orange) hemisphere. The position of every node is determined by a so-called force atlas, i.e. connected nodes are attracted to one another. Figure modified from [36]

erature and is related to the known asymmetries in localization of psychological functions, such as the prevalence of language functions in the left brain hemisphere in humans. Moreover, this lateralization is known to differ between genders [18, 56] as well as age [12]. Note that while a weak but systematic asymmetry between the hemispheres, i.e., a difference between the strength of corresponding connections in the left and right hemisphere, has been observed in our data set in more than 20 percent of the connections – most prominently in the temporal cortex – we have not observed significant variation of the asymmetry with gender, age, or handedness. This can probably be attributed to the relatively small size of our sample that provides information about the general structure and asymmetry of the brain, but not about its subject-specific variability.

Each node corresponding to a brain region is modeled by the two-dimensional FitzHugh-Nagumo model, a paradigmatic model for spiking neurons, which we have introduced in Sect. 2.1.1. The dynamics of the network read:

$$
\epsilon \dot{u}_k = u_k - \frac{u_k^3}{3} - v_k
$$
$$
+ \sigma \sum_{j \in N_H} A_{kj} \left[B_{uu}(u_j - u_k) + B_{uv}(v_j - v_k) \right] \tag{6.1a}
$$
$$
+ \varsigma \sum_{j \notin N_H} A_{kj} \left[B_{uu}(u_j - u_k) + B_{uv}(v_j - v_k) \right],
$$
$$
\dot{v}_k = v_k + a
$$
$$
+ \sigma \sum_{j \in N_H} A_{kj} \left[B_{vu}(u_j - u_k) + B_{vv}(v_j - v_k) \right] \tag{6.1b}
$$
$$
+ \varsigma \sum_{j \notin N_H} A_{kj} \left[B_{vu}(u_j - u_k) + B_{vv}(v_j - v_k) \right],
$$

with $k \in N_H$ where N_H denotes either the set of nodes k belonging to the left (N_L) or the right (N_R) hemisphere, and $\epsilon = 0.05$ describes the timescale separation between fast activator variable or neuron membrane potential u and the slow inhibitor or recovery variable v [15]. Depending on the threshold parameter a, the FitzHugh-Nagumo model may exhibit excitable behavior ($|a| > 1$) or self-sustained oscillations ($|a| < 1$). We use the FitzHugh-Nagumo model in the oscillatory regime and thus fix the threshold parameter at $a = 0.5$ sufficiently far from the Hopf bifurcation point. The emerging dynamics for an isolated FitzHugh-Nagumo oscillator is displayed in Fig. 2.4b. The coupling within the hemispheres is given by the intra-hemispheric coupling strength σ while the coupling between the hemispheres is given by the inter-hemispheric coupling strength ς. For the clarity of the calculations we set both coupling strengths to be equal $\varsigma = \sigma$ in the first part of this Chapter. The interaction scheme between nodes is characterized by a rotational coupling scheme:

$$\mathbf{B} = \begin{pmatrix} B_{uu} & B_{uv} \\ B_{vu} & B_{vv} \end{pmatrix} = \begin{pmatrix} \cos\phi & \sin\phi \\ -\sin\phi & \cos\phi \end{pmatrix}, \tag{6.2}$$

with coupling phase $\phi = \frac{\pi}{2} - 0.1$, causing primarily a cross-coupling. In Sect. 4.1.2, this particular scheme has been shown to be crucial for the occurrence of chimera states in ring topologies as it reduces the stability of the synchronized state.

6.1.2   Unihemispheric Slow-Wave Sleep

Humans and many species of animals require a state of rest during each day, called *sleep*. Although, sleep is a widely spread behavior across mammal evolution, understanding of its fundamental purpose and mechanism remains incomplete [40]. Nevertheless, the sleep of a healthy subject can be subdivided into individual stages by means of, inter alia, electroencephalograms (EEG), which have been introduced in Sect. 6.1. These stages are alternating among each other multiple times during a single sleep period. In Table 6.1, the different sleep stages are shown: In general, we can distinguish between rapid-eye-movement (R or REM) and non-rapid-eye-movement (N or NREM) sleep. The latter one can be further divided into three cases N1, N2, N3, whereby N3 is often referred to as deep sleep or slow-wave sleep. The single sleep stages can be characterized by means of main electroencephalograms (EEG) waves, in particular, by the bandwidth of the electroencephalograms-signals.

In the right column of Table 6.1, the timeseries of the electroencephalograms for the different sleep stages including wakefulness are depicted. Of particular interest are the slow waves in the δ-regime, which have the highest amplitude. As mentioned above, they are characteristic for the sleep stage called slow-wave sleep. Furthermore, they indicate a certain degree of synchronization of neural activity and imply synchronized firing in large neuronal populations [13, 48, 52]. For a more detailed description of sleep stages, we refer to [24, 46].

Among some birds and mammals like aquatic species (e.g., dolphins), unihemispheric slow-wave sleep is common [25, 30, 31, 38, 55]. This state can be characterized by dynamical asymmetry of the two hemispheres: One hemisphere of the brain exhibits a δ-wave dynamic, while the other remains in the α or β regime. This corresponds to a simultaneous but spatially separated occurrence of high and low degree of synchronization. One part of the brain is sleeping, while the other is awake.

Even though, unihemispheric slow-wave sleep has not been observed in the human brain, a similar phenomenon implies the ability of humans to exhibit dynamical asymmetries in the brain. Recently, the first-night effect has been explained by an interhemispheric asymmetry [54]. The first-night effect denotes the increased vigilance of one hemisphere, when sleeping in unfamiliar surroundings, and probably serves as a protection mechanism during the first night at a new place. Both phenomena, the unihemispheric slow-wave sleep as well as the first-night effect, can be explained in terms of partial synchronization.

Table 6.1 Classification of sleep stages after [6, 24, 59]: In general, we can distinguish between rapid-eye-movement (R) and non-rapid-eye-movement (N1, N2, N3) sleep, whereby, N3 is often referred to as deep sleep or slow-wave sleep. The single sleep stages can be characterized by means of main electroencephalograms (EEG) waves. The right column shows exemplary timeseries of the electroencephalograms for the different waves of the electroencephalography (EEG), taken from [1]

Waves		Frequency (in Hz)	(Sleep) stages					EEG
			Awake	R	N1	N2	N3	
Beta	β	15 - 25	•		•			
Alpha	α	8 - 12	•	•				
Theta	θ	4 - 7		•	•	•		
Delta	δ	0.5 - 4					•	

1 sec

6.2 Asymmetries in Hemispheric Networks

In the previous Sect. 6.1, we have introduced empirical structural brain networks and presented a model for dynamics on such networks. Our aim is to explore ways and means for partial synchronization patterns in hemispheric brain structures. These dynamical asymmetries can help to explain effects like unihemispheric slow-wave sleep or interhemispheric asymmetries. In this Section, we propose several measures for asymmetries in hemispheric networks and, by means of them, detect structural and dynamical asymmetries.

6.2.1 Measures for Asymmetries in Hemispheric Networks

Hemispheric mean phase velocity
We explore the dynamical behavior by calculating the mean phase velocity $\omega_k = 2\pi S_k/\Delta T$ for each node k, where ΔT denotes the time interval during which S_k complete rotations have been realized by the oscillator with node index k (see Sect. 4.1.1). Throughout this Chapter we use $\Delta T = 5000$. Furthermore we introduce hemispheric measures that characterize the degree of synchronization of the sub-networks and give complementary information. First, the hemispheric mean phase velocity is:

$$\langle \omega \rangle_H = \frac{1}{45} \sum_{k \in N_H} \omega_k, \tag{6.3}$$

where H denotes either the left ($H = L$) or right ($H = R$) hemisphere. Thus $\langle \omega \rangle_H$ corresponds to the mean phase velocity averaged over the left or right hemisphere, respectively. To quantify the dynamical difference between the left and right hemisphere we use the difference between these hemispheric mean phase velocities $\Delta \omega = \langle \omega \rangle_R - \langle \omega \rangle_L$.

Kuramoto order parameter
Second, we make use of the (Kuramoto) order parameter which we have introduced in Sect. 4.1.1. More precisely, we define the hemispheric Kuramoto order parameter as

$$R_H(t) = \frac{1}{45} \left| \sum_{k \in N_H} e^{i\theta_k(t)} \right|, \tag{6.4}$$

which is calculated by means of an abstract dynamical phase θ_k that can be obtained from the standard geometric phase $\tilde{\phi}_k(t) = \arctan(v_k/u_k)$ by a transformation which yields constant phase velocity $\dot{\theta}_k$. For an uncoupled FitzHugh-Nagumo oscillator the function $t(\tilde{\phi}_k)$ is calculated numerically, assigning a value of time $0 < t(\tilde{\phi}_k) < T$

for every value of the geometric phase, where T_P is the oscillation period. The dynamical phase is then defined as $\theta_k = 2\pi t(\tilde{\phi}_k)/T_P$, which yields $\dot{\theta}_k = $ const. Thereby identical, uncoupled oscillators have a constant phase relation with respect to the dynamical phase. Fluctuations of the order parameter R_H caused by the slow-fast time scales of the FitzHugh-Nagumo model are suppressed and a change in R_H indeed reflects a change in the degree of synchronization. The (Kuramoto) order parameter may vary between 0 and 1, where $R_H = 1$ corresponds to complete in-phase synchronization, and small values characterize spatially desynchronized states.

Spatial correlation coefficient

Finally, we use the spatial correlation coefficient, which has been proposed recently [20] to classify and distinguish between three types of chimera states: stationary, turbulent, and breathing chimeras. This coefficient is taking the pairwise distances $\{D_{kj}\}$ between the states of all oscillators k, j into account:

$$\{D_{kj}\} = \left| e^{i\theta_j} - e^{i\theta_k} \right|. \tag{6.5}$$

This distance is calculated using the dynamical phase on the unit cycle where the maximum distance of two oscillators is $D_{max} = 2$. The spatial correlation coefficient $g_0(t)$ generalizes the local curvature in systems with a spatial dimension and thereby measures the relative amount of synchronized oscillators. It is given by

$$g_0(t) = \sqrt{\int_0^{\delta} g(t, D)\mathrm{d}D}, \tag{6.6}$$

where $g(t, D)$ is the normalized probability density function, which is calculated as the probability of finding a distance D among all pairwise distances $\{D_{kj}\}$ in Eq. (6.5) at time t. For complete in-phase synchronization the distance between each pair of oscillators vanishes, i.e., $D = 0$ and $g(t, D) = \delta(D)$, hence $g_0(t) = 1$, while a totally incoherent system gives a value of $g(t, 0) = 0$, hence $g_0(t)$ is small. Two oscillators are considered spatially correlated if their distance is smaller than some threshold $\delta = 0.01 D_{max}$. The square root in Eq. (6.6) arises because by taking all pairwise distances, the probability of oscillators k and j both being in the synchronous cluster is proportional to the square of the number of synchronous oscillators. For a more detailed description we refer to [20].

6.2.2 Structural and Dynamical Asymmetries

Partial synchronization patterns as chimera states often occur in a regime, where in-phase synchronization is possible. The stability of in-phase synchronization in a network can be determined by means of the master stability function, introduced in Sect. 2.2.4. In contrast, the existence of frequency synchronization in a network has to be determined numerically as there exists no equivalent formalism. In Fig. 6.2, the regions of incoherence, frequency and in-phase synchronization is shown in the

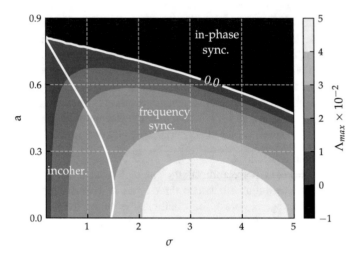

Fig. 6.2 Maximum Lyapunov exponent Λ_{max} in the parameter plane of the coupling strength σ and threshold parameter a: the three dynamical regimes of incoherence, frequency and in-phase synchronization are discriminated by white lines. The transition line between frequency and in-phase synchronization is calculated semi-analytically by means of the master stability function. The white line between these regimes indicates $\Lambda_{max} = 0$. The transition line between incoherence and frequency synchronization is determined fully numerically and is scored if at least 3 out of 5 numerical simulations of Eq. (6.1) end up in a frequency synchronized state. As initial conditions values for (u_k, v_k) from a circle with radius 2 and random phase are chosen. Other parameters are given by $\epsilon = 0.05$, $\phi = \frac{\pi}{2} - 0.1$. Figure modified from [35]

parameter plane of the coupling strength σ and threshold parameter a. The transition to in-phase synchronization is calculated by means of the master stability function, whereas the transition between incoherence and frequency synchronization is determined numerically. Further, due to bistability in the frequency synchronized regime as in [9], we require once frequency synchronized dynamics to remain frequency synchronized as the coupling strength increases, until eventually becoming in-phase synchronized.

In general, the transitions between the regimes in Fig. 6.2 differ slightly for the separate adjacency matrices $\{A_{kj}\}$ of the right $(k, j \in \{1, ..., 45\})$ and left $(k, j \in \{46, ..., 90\})$ hemisphere. Therefore, the transition lines in Fig. 6.2 can give us the information about possible parameters for partial synchronization patterns. For an appropriate coupling strength σ and threshold parameter a next to the transition lines, dynamical asymmetries become possible.

Thus, we investigate dynamical asymmetries emerging from the slight structural asymmetry of the brain hemispheres. Figure 6.3 shows how the different measures lead to the observation of a dynamical asymmetry with respect to the hemisphere of the average empirical structural brain network. Figure 6.3a displays the node-wise mean phase velocity ω_k for an intermediate coupling strength $\sigma = \varsigma = 0.7$ with random initial conditions. Note that the oscillators split into two visually well distinguishable communities that coincide with the hemispheres of the brain net

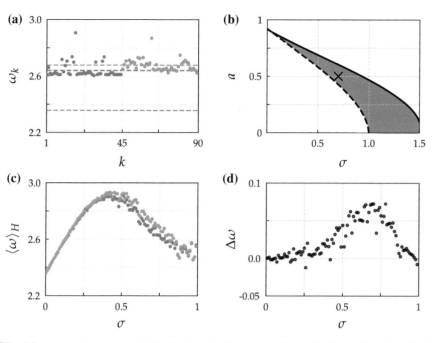

Fig. 6.3 Asymmetry scenarios in the brain network: **a** mean phase velocity ω_k for each node k (dots) and spatially averaged hemispheric mean phase velocity $\langle\omega\rangle_H$ (dashed lines) for coupling strength $\sigma = \varsigma = 0.7$. The color code highlights the left (dark blue) and right (light orange) hemisphere. The gray dashed line at $\omega \approx 2.4$ denotes the mean phase velocity for the uncoupled system. **b** Numerically calculated critical coupling strength in the (σ, a) plane for the transition between incoherence and frequency synchronization using the average brain network (solid line, $\sigma = \varsigma$ as in Fig. 6.2) and isolated hemispheres (dashed line, $\varsigma = 0$). The cross denotes the parameters ($\sigma = 0.7$, $a = 0.5$) used in panel (**a**). **c** Hemispheric mean phase velocities $\langle\omega\rangle_H$ as a function of the coupling strength $\sigma = \varsigma$, color code as in panel (**a**). **d** Difference between left and right hemispheric mean phase velocity $\Delta\omega$ as a function of the coupling strength $\sigma = \varsigma$. The difference assumes a maximum at $\sigma \approx 0.7$. Other parameters: $\epsilon = 0.05$, $a = 0.5$, $\phi = \frac{\pi}{2} - 0.1$. Figure from [36]

work and have different hemispheric mean phase velocities $\langle\omega\rangle_H$. The left and right hemispheric mean phase velocities $\langle\omega\rangle_L$ and $\langle\omega\rangle_R$ and their difference $\Delta\omega$ versus $\sigma = \varsigma$ are displayed in Fig. 6.3c, d, respectively. The values are calculated for one hundred different coupling strengths with $0 < \sigma \leq 1$ and step-size $\Delta\sigma = 0.01$. For every coupling strength an average over ten simulations with different sets of random initial conditions is plotted. For coupling strength $\sigma = \varsigma \geq 1$ the system enters the frequency-synchronized regime, while phase-synchronization measured by the Kuramoto order parameter sets in only later at $\sigma = \varsigma \approx 4.85$. Note that frequency synchronization means that all mean phase velocities ω_k are equal, but the phases may be different, while phase synchronization requires additionally that all phases are the same (see Sect. 2.2.4).

It turns out that the difference $\Delta\omega$ assumes a maximum at $\sigma \approx 0.7$ and subsequently decreases again as both hemispheres enter the frequency-synchronized regime. However, these differences between left and right hemisphere do not imply different dynamical regimes in the sense of a partial synchronization pattern consisting of a desynchronized and a synchronized hemisphere, like in a chimera pattern. Nevertheless it can clearly be concluded that the network dynamics reflects the slight structural asymmetry. Figure 6.3b depicts the critical coupling strength for the transition between incoherence and frequency synchronization for a wider range of parameters in the (σ, a) plane by a solid line for the coupled network with $\sigma = \varsigma$, and by a dashed line for the isolated hemispheres ($\varsigma = 0$). The additional coupling between the hemispheres leads to a higher threshold value σ_c for frequency synchronization.

So far, we have used an averaged empirical matrix to detect a dynamical asymmetry. For a deeper insight it is important to consider all twenty available empirical structural brain networks individually. In all, we observe one of three transition scenarios from incoherence ($0 < \sigma < 1$) to frequency synchronization ($\sigma > 1$ for $a = 0.5$) with increasing coupling strength, as shown in Fig. 6.4. They are distinguished by the difference of the hemispheric mean phase velocities $\Delta\omega$ exhibiting either a pronounced single maximum (Fig. 6.4d), or a (negative) minimum followed by a pronounced maximum (Fig. 6.4e), or essentially no dynamical asymmetry at all (Fig. 6.4f). However, in most cases (17 out of 20) a dynamical asymmetry has been measurable by means of $\Delta\omega$.

In the following, we analyze to which extent the dynamical asymmetry can be attributed to the structural asymmetry of the network by introducing a structural asymmetry parameter ρ with $0 \leq \rho \leq 1$ that allows for a continuous tuning between the original structural brain network and a fully symmetrized network, in the sense

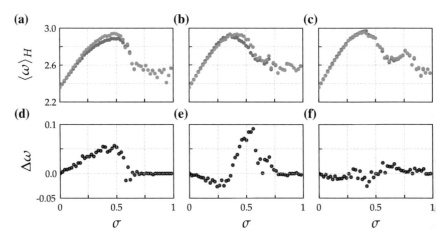

Fig. 6.4 Classification of the transition between incoherence and frequency synchronization by means of the hemispheric mean phase velocities $\langle\omega\rangle_H$ (**a**)–(**c**) and their difference $\Delta\omega$ (**d**)–(**f**) as a function of the coupling strength $\sigma = \varsigma$. In 20 individual brain networks three transition scenario are distinguished, displayed in panels (**a, d**), (**b, e**), and (**c, f**), respectively, each scored 10, 7, and 3 times, respectively. Other parameters as in Fig. 6.3. Figure from [36]

that both hemispheres are identical. We introduce the coupling matrix elements of a network interpolating between asymmetric and symmetric hemispheres by:

$$A^*_{kj} = \rho A_{kj} + (1 - \rho)\overline{A}_{kj}, \quad \rho \in [0, 1] \tag{6.7}$$

with $\overline{A}_{kj} = \frac{1}{2}\left(A_{kj} + A_{k+45,j+45}\right)$, where all indices are taken modulo 90. The resulting matrix $\{A^*_{kj}\}$ describes identical hemispheres if $\rho = 0$ and coincides with the original empirical matrix if $\rho = 1$.

We observe that the dynamical asymmetry, expressed by the hemispheric difference of the mean phase velocities $\Delta\omega$, builds up as the structural asymmetry parameter increases (Fig. 6.5a). The dynamical asymmetry is most pronounced for an intermediate coupling strength, not too small, but also not too close to the threshold of frequency synchronization. By integrating

$$W = \int_0^1 \Delta\omega \, d\sigma \tag{6.8}$$

we obtain a dynamical asymmetry parameter W which is indeed almost linearly correlated with the structural asymmetry parameter ρ. The Pearson correlation coefficient for the two measures is given by $r_{\rho,W} = 0.96$. The Pearson correlation coefficient $r_{x,y}$ is a standard measure to quantify the linear correlation between two samples of variables x and y:

$$r_{x,y} = \frac{\sum_{i=1}^{n}(x_i - \overline{x})(y_i - \overline{y})}{\sqrt{\sum_{i=1}^{n}(x_i - \overline{x})^2 \sum_{i=1}^{n}(y_i - \overline{y})^2}}, \tag{6.9}$$

where n is sample size, x_i, y_i are the individual sample points denoted by the index, and $\overline{x} = \frac{1}{n}\sum_{i=1}^{n} x_i$ is the sample mean (analogically for $\overline{y}$). A value of $r_{x,y} = 1$ means a total positive linear correlation, whereas $r_{x,y} = 0$ stands for no linear corre-

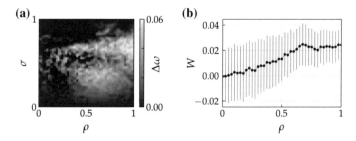

Fig. 6.5 Influence of structural asymmetry: **a** difference of the hemispheric mean phase velocities $\Delta\omega$ as a function of coupling strength $\sigma = \varsigma$ and structural asymmetry parameter ρ. The difference builds up as the structural asymmetry increases. **b** Dynamical asymmetry parameter W as a function of ρ. Bars denote standard deviation error of the mean with respect to 20 different realizations of the initial conditions. Other parameters as in Fig. 6.3. Figure modified from [36]

lation. Hence, the dynamical asymmetry parameter W increases linearly with ρ up to a certain degree of structural asymmetry, but then saturates and does not increase further (see Fig. 6.5b). This means that even the slightest structural asymmetry results already in a slight dynamical asymmetry, i.e., there is no threshold behavior. However, a slight dynamical asymmetry here does not induce an immediate symmetry breaking as known from critical phenomena. The increase of dynamical asymmetry instead firstly increases linearly with the structural asymmetry. Beyond this regime of linear response, the dynamical asymmetry does not increase further if the structural asymmetry increases beyond a certain degree, and the real empirical structural asymmetry seems to be just closely above that value corresponding to saturation of sensitivity. These results show consistency with our hypothesis that both unihemispheric sleep and bihemispheric sleeps can be possible dynamical states of the same network.

6.3   Partial Synchronization as a Model for Unihemispheric Sleep

In Sect. 6.1.2, we have described the phenomenon of unihemispheric sleep as state of partial synchronization. So far, we have used the empirical structural brain network derived from twenty healthy human subjects. Unihemispheric sleep is not known in case of humans, on the other hand, it has been reported that a variation of the coupling strength would support unihemispheric sleep [19]. Therefore, we will analyze the occurrence of partial synchronization in dependence on the inter-hemispheric coupling strength ς. Our motivation is based further on the hypothesis that sleeping with one hemisphere requires a degree of hemispheric separation [38]. It is conceivable that the corpus callosum – the main connection between the left and right hemisphere in the brain – may play an important role for unihemispheric sleep by controlling the coupling between the hemispheres.

6.3.1   Partial Synchronization Patterns

To achieve partial synchronization patterns, we consider the inter-hemispheric coupling strength ς as an independent parameter that allows us to reduce the coupling between the hemispheres. As mentioned above, this step is motivated by the presumption that sleeping with one hemisphere at a time requires a certain degree of hemispheric separation [38]. All other parameters remain unchanged.

We analyze the parameter regime where the previously used average empirical structural brain network with identical inter- and intra-hemispheric coupling strength $\varsigma = \sigma$ exhibits qualitatively different behavior from that with separated hemispheres $\varsigma = 0$, i.e., these two cases correspond to different dynamical regimes. For both cases we have numerically determined the critical intra-hemispheric coupling strength σ_c for which the system engages into the frequency synchronized regimes, see Fig. 6.3b. As $\varsigma = 0$ leaves us with two disconnected sub-networks, these sub-networks are naturally easier to synchronize. Note that these two disconnected sub-networks technically result in two different critical coupling strengths. However, the difference between these critical values is very small and thus negligible. Consider a coupling strength σ that lies within the shaded area of Fig. 6.3b. There, a phase transition with increasing ς must be expected, since the system is frequency synchronized if $\varsigma = 0$, and completely incoherent if $\varsigma = \sigma$.

We find that the frequency-synchronized solution indeed breaks down in one hemisphere. This gives rise to the partial synchronization pattern shown in Fig. 6.6 where the left hemisphere is incoherent while the right is frequency-synchronized, except for three small brain regions (hippocampus, gyrus parahippocampalis and amygdala). This shows up in the space-time plot, in the mean phase velocity profile, and in the hemispheric Kuramoto order parameter (although there is no perfect in-phase synchronization resulting in $R_R < 1$). Note that the incoherent, left hemisphere occasionally exhibits a high degree of synchronization that, in contrast to

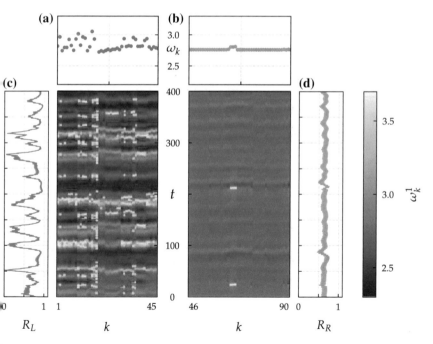

Fig. 6.6 Partial synchronization pattern for $\sigma = 0.70$, $\varsigma = 0.15$ with low and high degree of synchronization in the left (**a, c**) and right (**b, d**) hemisphere, respectively. **a, b** Mean phase velocity profiles ω_k. **c, d** inner panels: space-time plots of node-wise phase velocity ω_k^1 averaged over a single oscillation, outer panels: hemispheric Kuramoto order parameter $R_{L,R}$ as a function of time t. Other parameters as in Fig. 6.3. Figure from [36]

the right hemisphere, is unstable and vanishes after a short while. In general, partial synchronization patterns where different dynamical regimes occur in the two hemispheres can be found whenever a phase transition with respect to ς is expected, i.e. in the shaded region of Fig. 6.3b.

Finally, we analyze the transition from frequency to in-phase synchronization which occurs at much higher coupling strengths than shown in Fig. 6.3b, e.g., for $a = 0.5$ at $\sigma = \varsigma = 4.8$ and at $\sigma = 3.4$ for $\varsigma = 0$. Here we use the temporal mean of the spatial correlation coefficient g_0 that is suitable to distinguish between phase ($g_0 = 1$), frequency synchronization ($g_0 < 1$), and complete incoherence ($g_0 = 0$). In contrast to the Kuramoto order parameter R, the spatial correlation coefficient g_0 provides an arbitrary threshold δ and thereby gives a more pronounced transition from frequency to in-phase synchronization. Figure 6.7 shows that in the (σ, ς) plane of coupling strengths there exist regimes where a high degree of in-phase synchronization in one hemisphere coincides with a low degree of in-phase synchronization in the other (i.e., only frequency synchronization).

Further, we find that the degree of in-phase synchronization expressed by g_0 may exhibit non-monotonic behavior as a function of ς. To a certain amount this is expected as we have seen before that coupling two hemispheres ($\varsigma \neq 0$) decreases

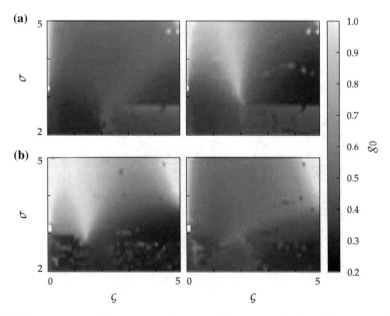

Fig. 6.7 Temporal mean of the spatial correlation coefficient g_0 as a function of the intra- and inter-layer coupling strengths σ, ς in the regime of strong coupling (phase or frequency synchronization). Left and right panels correspond to the left and right hemisphere, respectively. **a** and **b** depict the two possible states of the bistable system. The spatial correlation coefficient $g_0(t)$ is defined by Eq. (6.6), whereby $g_0 = 1$ stands for synchronization and $g_0 = 0$ for desynchronization. Other parameters as in Fig. 6.3. Figure from [36]

the degree of synchronization, we thus expect a maximum of g_0 at $\varsigma = 0$. However, in certain regimes of σ we find a subsequent maximum at $\varsigma \approx 2$ which implies that increasing the coupling between the hemispheres does not necessarily decrease the degree of in-phase synchronization. Furthermore, we find that this subsequent maximum is in principle possible in both hemispheres (indicating bistability, see Fig. 6.7a and b), but is realized by only one hemisphere at a time. A high degree of in-phase synchronization in one hemisphere thus suppresses in-phase synchronization in the other. This could be an important mechanism for the occurrence of unihemispheric sleep and should be investigated further. It is interesting that the two hemispheres can exchange their roles as one being phase synchronized and the other being only frequency-synchronized, depending upon the initial conditions. For very strong σ and ς complete in-phase synchronization of both hemispheres is found (top right corners in Fig. 6.7).

6.4 Summary

In this Chapter, we have investigated the dynamical asymmetry arising from the structural difference between the two brain hemispheres. We have taken into account empirical structural brain networks of human brains. Our aim has been to explore possibilities for the existence of unihemispheric sleep in hemispheric brain structures, where one hemisphere is in a slow-wave sleep stage while the other remains alert. Therefore, we have explained medical diagnosis methods such as electroencephalography (EEG) and various imaging procedures for the brain (Diffusion Tensor Imaging, Functional Magnetic Resonance Imaging, and diffusion-weighted Magnetic Resonance Imaging). It has been found that during the transition from complete incoherence to frequency synchronization an asymmetry regarding the degree of synchronization builds up, which can be quantified by the different mean phase velocities averaged over each hemisphere. We have shown that this asymmetry can be attributed to the structural asymmetry of the hemispheres, by introducing an asymmetry parameter which can interpolate between the empirical asymmetric brain network and an artificially symmetrized network.

Furthermore, we have varied the inter-hemispheric coupling strength, while keeping the intra-hemispheric coupling strength fixed, to increase the degree of inter-hemispheric separation, ranging from isolated to fully coupled hemispheres. This has resulted in the observation of partial synchronization patterns similar to spontaneously synchrony-breaking chimera states. We have demonstrated that these partial synchronization patterns occur for coupling strengths where the isolated hemispheres are frequency-synchronized while the brain network with equal intra- and inter-hemispheric coupling remains completely incoherent. By tuning the coupling between the hemispheres we have shown that at intermediate inter-hemispheric coupling one hemisphere becomes incoherent, giving rise to a chimera-like partial synchronization pattern.

These results are in accordance with the assumption that unihemispheric sleep requires a certain degree of inter-hemispheric separation. Moreover, it is known that the brain is operating in a critical state at the edge of different dynamical regimes [26], exhibiting hysteresis and avalanche phenomena as seen in critical phenomena and phase transitions [21, 41, 53]. By choosing appropriate coupling parameters, we have reported an intriguing dynamical behavior regarding the transition from frequency to in-phase synchronization. We observe that in this regime our brain model exhibits spontaneous symmetry breaking and bistability, where each hemisphere may engage into either of two dynamical states, characterized by a relatively high and low degree of synchronization. However, a high degree of synchronization in one of the hemispheres always coincides with a low degree of synchronization in the other. To sum up, the structural asymmetry in the brain allows for partial synchronization dynamics, which may be used to model unihemispheric sleep or explain the mechanism of the first-night effect in human sleep.

References

1. Abhang PA, Gawali BW (2015) Correlation of EEG images and speech signals for emotion analysis. Br J Appl Sci Technol **10**:1
2. Abrams DM, Strogatz SH (2004) Chimera states for coupled oscillators. Phys Rev Lett **93**:174102
3. Abrams DM, Mirollo RE, Strogatz SH, Wiley DA (2008) Solvable model for chimera states of coupled oscillators. Phys Rev Lett **101**:084103
4. Agnew HW Jr, Webb WB, Williams RL (1966) The first night effect: an EEG study of sleep. Psychophysiology **2**:263
5. Behrens TE, Berg HJ, Jbabdi S, Rushworth MF, Woolrich MW (2007) Probabilistic diffusion tractography with multiple fibre orientations: what can we gain? Neuroimage **34**:144–155
6. Bergner A, Frasca M, Sciuto G, Buscarino A, Ngamga EJ, Fortuna L, Kurths J (2012) Remote synchronization in star networks. Phys Rev E **85**:026208
7. Brown RE, Basheer R, McKenna JT, Strecker RE, McCarley RW (2012) Control of sleep and wakefulness. Physiol Rev **92**:1087
8. Cabral J, Fernandes HM, Van Hartevelt TJ, James AC, Kringelbach ML (2013) Structural connectivity in schizophrenia and its impact on the dynamics of spontaneous functional networks. Chaos **23**:046111
9. Chouzouris T, Omelchenko I, Zakharova A, Hlinka J, Jiruska P, Schöll E (2018) Chimera states in brain networks: empirical neural vs. modular fractal connectivity. Chaos **28**:045112
10. Costa MS, Born J, Claussen JC, Martinetz T (2016) Modeling the effect of sleep regulation on a neural mass model. J Comp Neurosci **41**:15
11. Creutzfeldt OD (1964) Experimenteller Nachweis von Beziehungen zwischen EEG-Wellen und Aktivität corticaler Nervenzellen. Naturwissenschaften **51**(7):166
12. Daianu M, Jahanshad N, Dennis EL, Toga AW, McMahon KL, de Zubicaray GI, Martin NG, Wright MJ, Hickie IB, Thompson PM (2012) Left versus right hemisphere differences in brain connectivity: 4-Tesla HARDI tractography in 569 twins. Proc IEEE Int Symp Biomed Imaging 526–529
13. Dang-Vu TT, Schabus M, Desseilles M, Albouy G, Boly M, Darsaud A, Gais S, Rauchs G, Sterpenich V, Vandewalle G, Carrier J, Moonen G, Balteau E, Degueldre C, Luxen A, Phillips C, Maquet P (2008) Spontaneous neural activity during human slow wave sleep. Proc Nat Acad Sci USA **105**:15160–15165

14. De Domenico M, Porter MA, Arenas A (2015) MuxViz: a tool for multilayer analysis and visualization of networks. J Complex Netw **3**:159–176
15. FitzHugh R (1961) Impulses and physiological states in theoretical models of nerve membrane. Biophys J **1**:445–466
16. Fuller PM, Gooley JJ, Saper CB (2006) Neurobiology of the sleep-wake cycle: sleep architecture, circadian regulation, and regulatory feedback. J Biol Rhythms **21**:482
17. Hlinka J, Coombes S (2012) Using computational models to relate structural and functional brain connectivity. Eur J Neurosc **36**:2137
18. Kann S, Zhang S, Manza P, Leung H-C, Li C-SR (2016) Hemispheric lateralization of resting-state functional connectivity of the anterior insula: association with age, gender, and a novelty-seeking trait. Brain Connect **6**(9):724–734
19. Kedziora DJ, Abeysuriya RG, Phillips AJK, Robinson PA (2012) Physiologically based quantitative modelling of unihemispheric sleep. J Theor Biol **314**:109
20. Kemeth FP, Haugland SW, Schmidt L, Kevrekidis YG, Krischer K (2016) A classification scheme for chimera states. Chaos **26**:094815
21. Kim H, Moon J-Y, Mashour GA, Lee U (2018) Mechanisms of hysteresis in human brain networks during transitions of consciousness and unconsciousness: theoretical principles and empirical evidence. PLoS Comput Biol **14**:e1006424
22. Kuramoto Y, Battogtokh D (2002) Coexistence of coherence and incoherence in nonlocally coupled phase oscillators. Nonlin Phen in Complex Sys **5**:380–385
23. Lesku JA, Vyssotski AL, Martinez-Gonzalez D, Wilzeck C, Rattenborg NC (2011) Local sleep homeostasis in the avian brain: convergence of sleep function in mammals and birds? Proc R Soc B **278**:2419
24. Malhotra RK, Avidan AY (2014) Sleep stages and scoring technique, Chapter 3. W.B. Saunders, pp 77–99
25. Mascetti GG (2016) Unihemispheric sleep and asymmetrical sleep: behavioral, neurophysiological, and functional perspectives. Nat Sci Sleep **8**:221
26. Massobrio P, de Arcangelis L, Pasquale V, Jensen HJ, Plenz D (2015) Criticality as a signature of healthy neural systems. Front Syst Neurosci **9**:22
27. Melicher T, Horacek J, Hlinka J, Spaniel F, Tintera J, Ibrahim I, Mikolas P, Novak T, Mohr P, Hoschl C (2015) White matter changes in first episode psychosis and their relation to the size of sample studied: a DTI study. Schizophr Res **162**:22–28
28. Moroni F, Nobili L, De Carli F, Massimini M, Francione S, Marzano C, Proserpio P, Cipolli C, De Gennaro L, Ferrara M (2012) Slow EEG rhythms and inter-hemispheric synchronization across sleep and wakefulness in the human hippocampus. NeuroImage **60**:497
29. Motter AE (2010) Nonlinear dynamics: spontaneous synchrony breaking. Nat Phys **6**:164–165
30. Mukhametov LM, Supin AY, Polyakova IG (1977) Interhemispheric asymmetry of the electroencephalographic sleep patterns in dolphins. Brain Res **134**:581
31. Niedernostheide FJ, Arps M, Dohmen R, Willebrand H, Purwins HG (1992) Spatial and spatio-temporal patterns in pnpn semiconductor devices. Phys Status Solidi (b) **172**:249
32. Olbrich E, Claussen JC, Achermann P (2011) The multiple time scales of sleep dynamics as a challenge for modelling the sleeping brain. Phil Trans R Soc A **369**:3884
33. Panaggio MJ, Abrams DM (2015) Chimera states: coexistence of coherence and incoherence in networks of coupled oscillators. Nonlinearity **28**:R67
34. Purpura DP (1959) Nature of electrocortical potentials and synaptic organizations in cerebral and cerebellar cortex. Int Rev Neurobiol **1**:47
35. Ramlow L (2018) Partial synchronization in 2-community networks of FitzHugh-Nagumo oscillators with empirical structural connectivities. Master's thesis, Technische Universität Berlin)
36. Ramlow L, Sawicki J, Zakharova A, Hlinka J, Claussen JC, Schöll E (2019) Partial synchronization in empirical brain networks as a model for unihemispheric sleep. Europhys Lett **126**:50007
37. Rattenborg NC, Lima SL, Amlaner CJ (1999) Facultative control of avian unihemispheric sleep under the risk of predation. Behav Brain Res **105**:163

38. Rattenborg NC, Amlaner CJ, Lima SL (2000) Behavioral, neurophysiological and evolutionary perspectives on unihemispheric sleep. Neurosci Biobehav Rev **24**:817–842
39. Rattenborg NC, Voirin B, Cruz SM, Tisdale R, Dell'Omo G, Lipp HP, Wikelski M, Vyssotski AL (2016) Evidence that birds sleep in mid-flight. Nat Commun **7**:12468
40. Rattenborg NC, Horacio O, Kempenaers B, Lesku JA, Meerlo P, Scriba MF (2017) Sleep research goes wild: new methods and approaches to investigate the ecology, evolution and functions of sleep. Phil Trans R Soc B **372**:20160251
41. Ribeiro TL, Copelli M, Caixeta F, Belchior H, Chialvo DR, Nicolelis MAL, Ribeiro S (2010) Spike Avalanches exhibit universal dynamics across the sleep-wake cycle. PLoS ONE **5**:e14129
42. Saper CB, Chou TC, Scammell TE (2001) The sleep switch: hypothalamic control of sleep and wakefulness. Trends Neurosci **24**:726
43. Saper CB, Fuller PM, Pedersen NP, Lu J, Scammell TE (2010) Sleep state switching. Neuron **68**:1023
44. Scammell TE, Arrigoni E, Lipton JO (2017) Neural circuitry of wakefulness and sleep. Neuron **93**:747
45. Schilling KG, Daducci A, Maier-Hein K, Poupon C, Houde J-C, Nath V, Anderson AW, Landman BA, Descoteaux M (2019) Challenges in diffusion MRI tractography—lessons learned from international benchmark competitions. Magn Res Imaging **57**:194
46. Schwartz JRL, Roth T (2008) Neurophysiology of sleep and wakefulness: basic science and clinical implications. Curr Neuropharmacol **6**:367–378
47. Schöll E (2016) Synchronization patterns and chimera states in complex networks: interplay of topology and dynamics. Eur Phys J Spec Top **225**:891–919
48. Sejnowski TJ, Destexhe A (2000) Why do we sleep? Brain Res **886**:208–223
49. Shima S, Kuramoto Y (2004) Rotating spiral waves with phase-randomized core in nonlocally coupled oscillators. Phys Rev E **69**:036213
50. Soriano MC, García-Ojalvo J, Mirasso CR, Fischer I (2013) Complex photonics: dynamics and applications of delay-coupled semiconductors lasers. Rev Mod Phys **85**:421–470
51. Spiess M, Bernard G, Kurth S, Ringli M, Wehrle FM, Jenni OG, Huber R, Siclari F (2018) How do children fall asleep? A high-density EEG study of slow waves in the transition from wake to sleep. NeuroImage **178**:23
52. Steriade M, McCormick DA, Sejnowski TJ (1993) Thalamocortical oscillations in the sleeping and aroused brain. Science **262**:679–685
53. Steyn-Ross DA, Steyn-Ross M (2010) Modeling phase transitions in the brain. Springer, New York
54. Tamaki M, Bang JW, Watanabe T, Sasaki Y (2016) Night watch in one brain hemisphere during sleep associated with the first-night effect in humans. Curr Biol **26**:1190–1194
55. Tarpley RJ, Ridgway SH (1994) Corpus callosum size in delphinid cetaceans. Brain Behav Evol **44**(3):156
56. Tomasi D, Volkow ND (2012) Laterality patterns of brain functional connectivity: gender effects. Cereb Cortex **22**(6):1455–1462
57. Tzourio-Mazoyer N, Landeau B, Papathanassiou D, Crivello F, Etard O, Delcroix N, Mazoyer B, Joliot M (2002) Automated anatomical labeling of activations in SPM using a macroscopic anatomical parcellation of the MNI MRI single-subject brain. Neuroimage **15**:273–289
58. Vyazovskiy VV, Olcese U, Hanlon EC, Nir Y, Cirelli C, Tononi G (2011) Local sleep in awake rats. Nature **472**:443
59. Wright KP (2009) Encyclopedia of neuroscience, Chapter EEG in S. Springer, p 85

Chapter 7
Multiplex Networks

Network science presents a unique platform to study various complex real-world systems by analyzing the interactions between their constituent entities and collectively investigating their behaviors [3, 38, 39]. In the last Chap. 6, we have studied partial synchronization in a 2-community network, which represents the two hemispheres of the human brain. A recent addition to the network science is the multiplex framework (see Sect. 2.2.2) which incorporates multiple types of interactions among nodes by representing them in different layers [12, 55]. For example, the neurons in the brain form different groups consisting of the same neurons but interacting in different ways (chemical interaction or electrical synapses) to perform different tasks [20, 44]. Multiplex framework divides these neuronal groups into different layers based on the functionalities of the groups [23]. Similarly, transportation networks, communication networks, social networks and a lot of other real-world networks can be represented in a multiplex framework to understand their structural and dynamical features in a better fashion [41].

Moreover, complex networks consisting of several interacting layers allow for remote synchronization of distant layers via an intermediate relay layer, as has been shown recently [43]. In such case there exist synchronization between one layer and a second layer, where these two layers are not directly connected. A simple realization of such a system is a three-layer multiplex network where a relay layer in the middle, which is generally not synchronized, acts as a transmitter between two outer layers. In the nervous system relaying information between specific neural populations is highly general and necessary [21, 36]. During this relay mechanism the information transmission is delayed in between separate areas relatively to neural activity transmission within a single spatially confined population. Such interlayer delay in multiplex networks might play as a general organizing force to synchronize spatiotemporal patterns between layers. It is the purpose of the present Chapter to extend the notion of relay synchronization from completely synchronized states to partial synchronization patterns in the individual layers and study various scenarios of

© Springer Nature Switzerland AG 2019
. Sawicki, *Delay Controlled Partial Synchronization in Complex Networks*,
Springer Theses, https://doi.org/10.1007/978-3-030-34076-6_7

synchronization of chimera states in a triplex network. This Chapter includes contents that have been published in [56–58, 75, 76] (for [57]: © 2018 by the American Physical Society (APS)).

The outline of the Chapter is as follows: In Sect. 7.1, we extend the notion of relay synchronization to chimera states, and study the scenarios of relay synchronization in a three-layer network of FitzHugh-Nagumo oscillators, where each layer has a nonlocal coupling topology. In Sect. 7.2, we establish time delay in the inter- and intra-layer coupling as a powerful tool to control various synchronization patterns, in particular chimera states. Our analysis in Sect. 7.3 shows that the three-layer structure of the network gives also rise to partial synchronization of chimera states in the outer layers via the relay layer. We demonstrate that the three-layer structure of the network allows for synchronization of the coherent domains of chimera states in the first layer with their counterparts in the third layer, whereas the incoherent domains remain desynchronized. We conclude with Sect. 7.6 summarizing our results.

7.1 Relay Synchronization

In the present Section, we study scenarios of remote synchronization for chimera states in a three-layer multiplex network of FitzHugh-Nagumo oscillators. This model is a paradigmatic system widely used in neuroscience and electrical engineering (see Sect. 2.1.1). This Section closely follows [56, 57] [for [57]: © 2018 by the American Physical Society (APS)]. We demonstrate that the three-layer structure of the network allows for synchronization of chimera states in the outer layers via the relay layer. We analyze the influence of the inter-layer coupling upon the relay synchronization of chimera states. As a useful measure for relay synchronization, we introduce the global inter-layer synchronization error in Sect. 7.1.2.

7.1.1 Triplex Network

We consider a multiplex network with $M = 3$ layers (triplex) as shown in Fig. 7.1. Each layer consists of a ring of N identical FitzHugh-Nagumo oscillators with non-

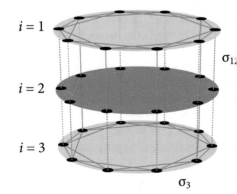

Fig. 7.1 Schematic triplex network with ordinal coupling. The middle layer $i = 2$ (red) acts as relay layer between the two outer layers $i = 1, 3$ (blue). We distinguish between intra-layer (full lines), e.g. σ_3, and inter-layer (dashed lines) couplings, e.g. σ_{12}, between the nodes (black dots). Figure modified from [56]

$i = 1$

$i = 2$

$i = 3$

σ_{12}

σ_3

local (intra-layer) coupling. Layers $i = 1$ and $i = 3$ are coupled with the intermediate layer $i = 2$ (ordinal inter-layer coupling, see Sect. 2.2.2), so that the middle layer acts as a relay layer between the two outer layers, but there is no inter-layer coupling between layers 1 and 3. The dynamical equations for the activator (u) and inhibitor (v) variables $\mathbf{x} = (u, v)^T \in \mathbb{R}^2$ are:

$$\dot{\mathbf{x}}_k^i(t) = \underbrace{\mathbf{F}(\mathbf{x}_k^i(t))}_{\text{local dynamics}} + \underbrace{\frac{\sigma_i}{2R_i} \sum_{l=k-R_i}^{k+R_i} \mathbf{H}[\mathbf{x}_l^i(t) - \mathbf{x}_k^i(t)]}_{\text{intra-layer coupling}} + \underbrace{\sum_{j=1}^{M=3} \sigma_{ij} \mathbf{H}[\mathbf{x}_k^j(t) - \mathbf{x}_k^i(t)]}_{\text{inter-layer coupling}}$$

(7.1)

with $k \in \{1, \ldots, N\}$, $i \in \{1, \ldots, 3\}$; all indices k, l are modulo N. The coupling radius in layer i is denoted by R_i. The dynamics of each individual oscillator is governed by

$$\mathbf{F}(\mathbf{x}) = \begin{pmatrix} \varepsilon^{-1}(u - \frac{u^3}{3} - v) \\ u + a \end{pmatrix},$$

(7.2)

where $\varepsilon > 0$ is a small parameter characterizing the time scale separation, which we fix at $\varepsilon = 0.05$ throughout the current Chapter. Depending on the threshold parameter a the FitzHugh-Nagumo element exhibits either oscillatory ($|a| < 1$) or excitable ($|a| > 1$) behavior. We consider the oscillatory regime ($a = 0.5$) in this Chapter. The parameter σ_i denotes the coupling strength inside the layer (intra-layer coupling), and σ_{ij} is the inter-layer coupling. In order to ensure constant row sum we choose the inter-layer coupling as $\sigma_{12} = \sigma_{23}$, which gives the inter-layer coupling matrix for an ordinal inter-layer coupling (see Sect. 2.2.2)

$$\sigma = \begin{pmatrix} 0 & \sigma_{12} & 0 \\ \frac{\sigma_{12}}{2} & 0 & \frac{\sigma_{23}}{2} \\ 0 & \sigma_{23} & 0 \end{pmatrix}.$$

(7.3)

The interaction is realized through diffusive coupling with coupling matrix

$$\mathbf{H} = \begin{pmatrix} \varepsilon^{-1}\cos\phi & \varepsilon^{-1}\sin\phi \\ -\sin\phi & \cos\phi \end{pmatrix}$$

(7.4)

and coupling phase $\phi = \frac{\pi}{2} - 0.1$. This coupling configuration (i.e., predominantly activator-inhibitor cross-coupling) is similar to a phase-lag of approximately $\pi/2$ in the Kuramoto model and has been chosen such that chimeras are most likely (see Sect. 4.1.2).

7.1.2 Global Synchronization Error

An appropriate measure for instantaneous synchronization between two layers i, j is the global inter-layer synchronization error E^{ij}, defined by

$$E^{ij} = \lim_{T \to \infty} \frac{1}{NT} \int_0^T \sum_{k=1}^N \left\| \mathbf{x}_k^j(t) - \mathbf{x}_k^i(t) \right\| \, dt, \qquad (7.5$$

where $\|\cdot\|$ stands for the Euclidean norm, and the normalization by N allows for better comparison of networks of different size as shown in Sect. 7.1.3. First we consider a network with three identical layers with an ordinal inter-layer coupling (see Sect. 2.2.2). Such a network is able to demonstrate remote synchronization of distant layers via a relay layer. Regarding the global inter-layer synchronization three dynamical regimes are conceivable:

- *full inter-layer synchronization* where synchronization exists between all three layers ($E^{12} = E^{13} = 0$),
- *relay inter-layer synchronization* where synchronization exists just between the two outer layers ($E^{12} \neq 0$ and $E^{13} = 0$),
- *inter-layer desynchronization* ($E^{12} \neq E^{13} \neq 0$).

Therefore, by measuring the global inter-layer synchronization error between the first and second layer E^{12} and between the first and third layer E^{13}, we can distinguish between these synchronization mechanism.

For an appropriate coupling strength σ_i and coupling radius R_i, chimera states appear inside the ith layer. These complex patterns have been found the first time in a ring of oscillators which can be seen as a single layer system [42]. Numerical simulations in Fig. 7.2 show that we can observe all of the three dynamical regimes depending on the initial conditions (here: random initial conditions), whereby each layer exhibits a chimera state (see insets of Fig. 7.2): Already for small inter-layer coupling strength $\sigma_{ij} > 0.01$ relay inter-layer synchronization is possible. By increasing σ_{ij}, the middle layer also tends to synchronize with the outer layers. At a critical value of the coupling strength ($\sigma_{ij} = 0.12$) this relay synchronization mechanism cannot be observed anymore, and the full inter-layer synchronization mechanism remains. This mechanism exists over the whole range of σ_{ij}: By increasing σ_{ij} the three layers tend to full inter-layer synchronization at $\sigma_{ij} = 0.24$. For $\sigma_{ij} < 0.01$ no synchronization can be observed in case of random initial conditions. Nevertheless full inter-layer synchronization can always be achieved for any value of σ_{ij} by choosing full inter-layer synchronization as initial conditions. In that case the inter-layer coupling term in Eq. (7.1) vanishes due to the diffusive coupling form.

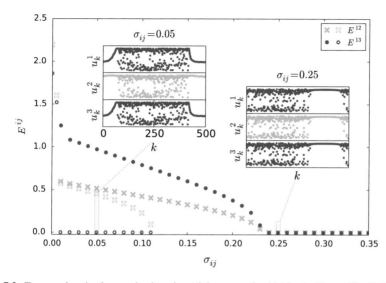

Fig. 7.2 Two synchronization mechanisms in a triplex network with identical layers Eq. (7.1): The main plot shows the global inter-layer synchronization error E^{ij} between the first and second layer yellow crosses, E^{12}) and between the outer layers (blue circles, E^{13}) versus the inter-layer coupling $\sigma_{ij} = \sigma_{12} = \sigma_{23}$. The filled markers correspond to the full synchronization mechanism, whereas the unfilled ones to the relay synchronization mechanism. Already for small inter-layer couplings (0.01 < σ_{ij} < 0.12) we can observe either relay inter-layer synchronization or the transition to full inter-layer synchronization, whereas for greater values of the coupling strength (σ_{ij} > 0.12) just the latter one is observable. The insets show snapshots of variables u_k^i for all three layers i on top of each other for two different inter-layer coupling strengths $\sigma_{ij} = 0.05$ and 0.25, respectively. These values correspond to the two synchronization mechanisms, respectively. For all simulations of the full Eq. (7.1) random initial conditions are used (different ones for the two mechanisms). Parameters are chosen as $\varepsilon = 0.05$, $a = 0.5$, $\sigma_i = 0.05$, $N = 500$, $R_i = 170$, $\phi = \frac{\pi}{2} - 0.1$, and = 1, 2, 3. Figure from [57]. © 2018 by the American Physical Society (APS)

7.1.3 Network Size Dependence

Similar quantifications to measure the inter-layer synchronization have been introduced recently [43]. In our definition of the global inter-layer synchronization error, the maximum value of E^{ij} does not depend on the node number N as we can see in Fig. 7.3: Compared to Fig. 7.2 the network size is doubled but the values of the global synchronization errors E^{ij} remain the same. The reason is the normalization by N in Eq. (7.5). A missing normalization would make a comparison between various network sizes more difficult. A further step could be to introduce a normalization to one for the purpose of comparison between different local dynamics.

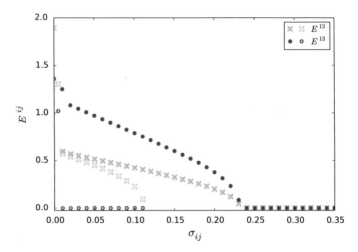

Fig. 7.3 Two synchronization mechanisms in a triplex network of size $N = 1000$: The plot shows the global inter-layer synchronization error E^{ij} between the first and second layer (yellow crosses, E^{12}) and between the outer layers (blue circles, E^{13}) versus the inter-layer coupling $\sigma_{ij} = \sigma_{12} = \sigma_{23}$. The filled markers correspond to the full synchronization mechanism, whereas the unfilled ones to the relay synchronization mechanism. For all simulations random initial conditions are used (different ones for the two mechanisms). The system has the double size compared to $N = 500$ in Fig. 7.2. Coupling range is given by $R_i = 340$, other parameters as in Fig. 7.2. Figure from [57] © 2018 by the American Physical Society (APS)

7.2 Influence of Delay

Our focus in this Section is on the control of the chimera synchronization pattern by time delay in the inter- and intra-layer coupling in a multi(tri)plex network. This Section is closely related to [56, 57] [for [57]: © 2018 by the American Physical Society (APS)]. As mentioned in Sect. 5.2, time delays represent an essential factor in real-world networks due to the finite speed of information propagating through channels connecting the nodes. They play a crucial role in determining the dynamical behavior of a complex system [2, 7, 8, 46, 61, 69, 70]. Time delays have been shown to heavily influence the parameter range for which chimera states appear for both single and multiplex networks [29, 31, 57, 63, 81]. Moreover, recently a scheme has been proposed for engineering chimera states in multilayer systems using suitably placed heterogeneous delays [30]. Introducing time delay in ring networks often leads to traveling patterns which makes it difficult to calculate observables averaged over time and space. Therefore, in Sect. 7.2.6, we introduce a detrending method to avoid these difficulties. For a single-layer network it is known that for appropriate coupling strength σ_i and coupling range R_i complex patterns of spatially coexisting coherent and incoherent dynamics, i.e., chimera states, can occur and they may be centered at different spatial locations depending on the initial conditions (see Chaps. 4 and 5). On the other hand, we have shown in the previous Sect. 7.1 that in multiplex networks one can achieve synchronization of either neighboring or remote layers. Here we establish the possibility to control synchronization patterns even of remote layers, in particular chimera states, by tuning the delay time in the coupling terms. Varying this parameter allows for an overall control of the dynamical regimes in the network.

7.2.1 Delay Between the Layers

First we introduce time delay τ only in the inter-layer coupling, since in real-world systems the transfer of information between two different layers is often slower than within one layer.

So the difference between the system considered in this Section and the previous Sect. 7.1.1 is an additional inter-layer delay time τ. The dynamical equations are given by

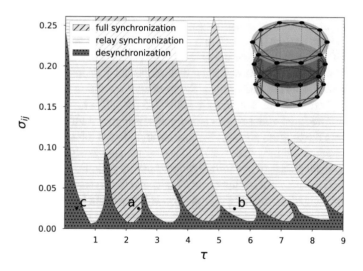

Fig. 7.4 Relay synchronization tongues in the parameter plane of inter-layer coupling strength $\sigma_{ij} \equiv \sigma_{12} = \sigma_{23}$ and inter-layer coupling delay $\tau > 0$: Inter-layer relay synchronization (horizontally hatched yellow region) occurs between regions of full inter-layer synchronization (diagonally hatched blue region) and desynchronized inter-layer dynamics (dotted dark red regions). Black dots (a, b, c) denote parameter values of the synchronization scenarios shown in Fig. 7.5. Random initial conditions are used for all numerical simulations. Parameters: $\varepsilon = 0.05$, $a = 0.5$, $\sigma_i = 0.2$, $R_i = 170$ for $i = 1, 2, 3$, $\phi = \frac{\pi}{2} - 0.1$, $N = 500$. The inset shows a schematic triplex network. The middle layer $i = 2$ (dark red) acts as relay layer between the outer layers $i = 1, 3$ (blue). Figure from [57]. © 2018 by the American Physical Society (APS)

$$\dot{\mathbf{x}}_k^i(t) = \underbrace{\mathbf{F}(\mathbf{x}_k^i(t))}_{\text{local dynamics}} + \underbrace{\frac{\sigma_i}{2R_i} \sum_{l=k-R_i}^{k+R_i} \mathbf{H}[\mathbf{x}_l^i(t) - \mathbf{x}_k^i(t)]}_{\text{intra-layer coupling}} + \underbrace{\sum_{j=1}^{3} \sigma_{ij} \mathbf{H}[\mathbf{x}_k^j(t-\tau) - \mathbf{x}_k^i(t)]}_{\text{inter-layer coupling}},$$

(7.6)

where all variables are defined as in Eq. (7.1).

Numerical simulations in Fig. 7.4 show that we can observe all aforementioned synchronization scenarios depending on the parameters and the initial conditions (here: random initial conditions). When the layers are coupled weakly, they tend to behave independently (red dotted region): Each layer exhibits a chimera state but there is no synchronization between the layers. With increasing delay $\tau > 0$ (the case $\tau = 0$ has been treated separately in Sect. 7.1.1) we observe a sequence of tongue-like regions in the parameter plane (τ, σ_{ij}): Full inter-layer synchronization (blue regions with diagonal stripes) alternating with relay inter-layer synchronization (yellow regions with horizontal stripes). Exemplary snapshots of the dynamics in these synchronized regions are shown in Fig. 7.5a, b (left column). The right column of Fig. 7.5 will be explained in the next Section. We can observe full in-phase synchronization of all three layers for values of τ close to integer multiples of the period of the uncoupled system $T = 2.3$, and relay inter-layer synchronization with anti-phase synchronization between the outer layers and the relay layer for half-integer multiples. Analytical calculations show that the period T decreases with increasing σ_{ij} (see Sect. 7.2.4). Therefore, due to the resonance condition of τ with respect to the intrinsic period T, the tongues are shifted to the left with increasing coupling strength σ_{ij}. The same effect occurs when τ equals higher multiples of the intrinsic period, where the tongues are shifted more strongly to the left and decrease in size, which is a general feature of resonance tongues in delay systems [34, 78].

In Fig. 7.6, the effects of the shifted tongues can be seen. For a fixed $\tau = 1.5$, the global inter-layer synchronization error E^{ij} between the layers is analyzed in dependence on the inter-layer coupling σ_{ij}. As expected, we obtain full synchronization by increasing the inter-layer coupling strength to $\sigma_{ij} \approx 0.05$. Depending on the intra-layer coupling σ_i, this transition is shifted to slightly higher values of the inter-layer coupling ($\sigma_{ij} = 0.06$ in Fig. 7.6d). It is remarkable that the intra-layer coupling has such a weak influence on the transition to full synchronization. Nevertheless, a further increase yields a transition from full to relay synchronization. In Fig. 7.6a, this transition occurs for $\sigma_{ij} = 0.25$ in a smooth way. For a stronger intra-layer coupling σ_i, the transition becomes sharper: In comparison to $\sigma_i = 0.10$ in Fig. 7.6a, a stronger intra-layer coupling $\sigma_i = 0.35$ in Fig. 7.6d leads to an abrupt transition at $\sigma_{ij} = 0.35$. We have explained analytically the curvature of the delay tongues in dependence on the intra-layer coupling in Sect. 5.2.3.

The mechanism behind it, is the modulation of the intrinsic period of the oscillator as shown in Eq. (5.14). Therefore, by an increase of the intra-layer coupling strength for a fixed delay time, the transition between the synchronization tongues happens much faster. On the one side, we find a slight increase in Fig. 7.6a, b and c of the error, on the other side a sudden transition in Fig. 7.6d can be measured. In contrast

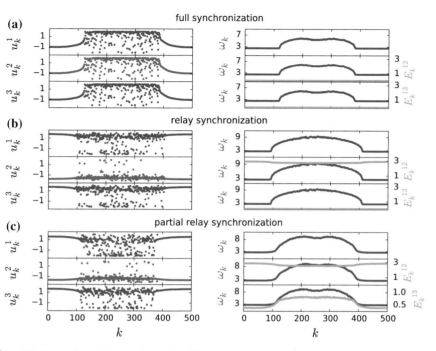

Fig. 7.5 Dynamics of the three layers for different values of delay time τ, marked by black circles (a,b,c) in Fig. 7.4: **a** Full inter-layer synchronization for $\tau = 2.4$. **b** Relay inter-layer synchronization for $\tau = 5.5$. **c** Partial relay inter-layer synchronization between the outer layers for $\tau = 0.4$. The left column shows snapshots of variables u_k^i for all three layers $i = 1, 2, 3$ (relay layer: light red, outer layers: dark blue), whereas the right column shows the corresponding mean phase velocity profiles ω_k (dark blue) for each layer and local inter-layer synchronization error E_k^{ij} (light yellow), introduced in the next Sect. 7.2.2. Inter-layer coupling is given by $\sigma_{ij} = 0.025$, other parameters as in Fig. 7.4. Figure from [57]. © 2018 by the American Physical Society (APS)

to the first transition to full synchronization for small inter-layer coupling $\sigma_{ij} \approx 0.05$, the transition from full to relay synchronization depends to a much greater degree on the intra-layer coupling σ_i. This complex interplay of inter- and intra-layer couplings is due to the complex dynamics within each layer, which have not been taken into account in recent publication [43]. For a deeper insight into the synchronization scenarios we will introduce a more accurate measure in the next Sect. 7.2.2.

7.2.2   Local Synchronization Error

In Sect. 7.1.2, we have introduced the global synchronization error E^{ij} in Eq. (7.5). This measure gives information about the mean synchronization degree between two layers, regardless of the intra-layer dynamic. To provide more insight into the synchronizability of patterns between two layers i, j, in particular the synchronization

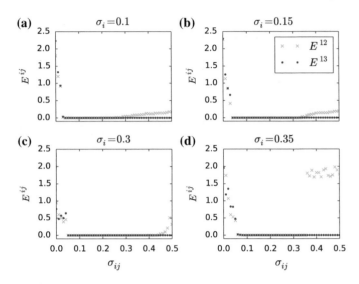

Fig. 7.6 Transition between full and relay synchronization in a triplex network with identical layers Eq. (7.6): The panels show the global inter-layer synchronization error E^{ij} between the first and second layer (light yellow crosses, E^{12}) and between the outer layers (dark blue circles, E^{13} versus the inter-layer coupling $\sigma_{ij} = \sigma_{12} = \sigma_{23}$ for four different values of the intra-layer coupling σ_i: **a** $\sigma_i = 0.1$, **b** $\sigma_i = 0.15$, **c** $\sigma_i = 0.3$, and **d** $\sigma_i = 0.35$. For all simulations of the full Eq. (7.6) random initial conditions are used. Other parameters are chosen as $\tau_{ij} = \tau_{12} = \tau_{23} = 1.5, \varepsilon = 0.05$ $a = 0.5, N = 500, R_i = 170, \phi = \frac{\pi}{2} - 0.1$, and $i = 1, 2, 3$

of chimera states, we use the local inter-layer synchronization error in dependence of every single node k:

$$E_k^{ij} = \lim_{T \to \infty} \frac{1}{T} \int_0^T \left\| \mathbf{x}_k^j(t) - \mathbf{x}_k^i(t) \right\| \mathrm{d}t, \qquad (7.7$$

where $\|\cdot\|$ stands for the Euclidean norm. The local inter-layer synchronization error is convenient for detecting those nodes which are synchronized between two layers especially in the (red dotted) region of desynchronization in Fig. 7.4. Exemplary dynamics inside this region are given in Fig. 7.5c and will be the focus of the next Sect. 7.3. In Fig. 7.5, the local synchronization error is plotted in the right column (light yellow). Such a synchronization scenario cannot be detected by the global inter-layer synchronization error $E^{ij} \neq 0$, but the node-dependent local measure E_k^i gives us the possibility to detect this type of synchronization.

7.2.3   Breathing Chimera States

In [40] general measures for chimera states have been presented. The authors distinguish between different scenarios of chimera states. Among them, breathing chimera states represent a particular scenario. In multiplex networks we are also able to observe breathing chimeras. In Fig. 7.7, the position of the coherent domain is alternating within each layer. In the space-time of the first layer, the center of the coherent part of the chimera state is at $k = 350$ for $t = 6500$. For $t = 10{,}000$ the center is at $k = 100$, where previously the incoherent domain has been. The chimera state is shifted exactly to the other side of the layer-ring ($350 - 100 = N/2$), which recently have been also observed in coherence resonance chimeras [64]. We find exactly the same breathing dynamics in the outer layers. In the relay layer two scenarios can be found: Either the relay layer behaves like the outer layers (this is the case in Fig. 7.7), or the position of coherent and incoherent domain in the relay layer can be interchanged, respectively. An exemplary snapshot of such an "anti"-synchronized state is depicted in Fig. 7.13a. We will give an analytical explanation for this mechanism in Sect. 7.5.

The period of the breathing depends on the delay time τ and the initial conditions. Moreover, we can observe breathing chimeras, which travel around the ring similar to traveling chimeras in a single-layer system as shown in Fig. 5.8. The breathing chimera state shown in Fig. 7.7 is also traveling: For $t \approx 6500$, the center of the coherent domain is located around $k \approx 325$, whereas for $t \approx 8500$, the center is slightly shifted to the right ($k \approx 350$). The same holds true for the other center at $k \approx 90$ for $t \approx 7500$, moving to $k \approx 115$ for $t \approx 10{,}000$. The speed of these traveling structures depends on the system parameter and the initial conditions. Cutting all multiplex links ($\sigma_{ij} = 0$), we obtain non-breathing chimera states in all layers. The multiplex structure ($\sigma_{ij} \neq 0$) destabilizes these chimera states once they have appeared. The destabilization yields a shrinking process over a time of $\Delta t > 1000$. As shown for the breathing chimeras in Fig. 7.7, the birth of a chimera state is much faster ($\Delta t < 100$) than its shrinking process ($\Delta t > 1000$).

7.2.4   Estimate of Period

In many delay systems one observes resonance effects if the delay is an integer or half-integer multiples of the period of the uncoupled system [34, 78, 79]. For full inter-layer synchronization the undelayed part of the coupling term in the ith layer is the most important part in case of incoherent dynamics and can be rewritten as follows, neglecting $\cos\phi \ll 1$ and setting $\sin\phi \approx 1$:

$$\varepsilon\dot{u} = u - \frac{u^3}{3} - (1 + \sigma_i + \sigma_{ij})v,$$
$$\dot{v} = (1 + \sigma_i + \sigma_{ij})u + a,$$

$$(7.8)$$

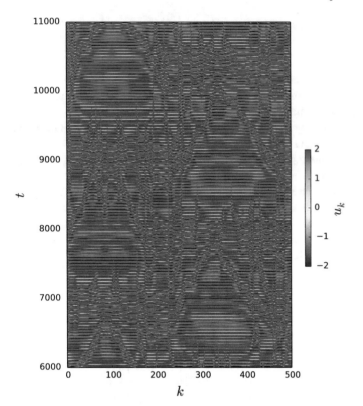

Fig. 7.7 Breathing chimera state in a triplex network with identical layers: Space-time plot of the variable u_k^1 (first layer) of breathing chimera states, where the position of the coherent and incoherent domains is alternating for $\tau = 1.7$, $\sigma_i = 0.1$ and $\sigma_{ij} = 0.04$. All layers are fully synchronized in this simulation ($u_k^1 = u_k^2 = u_k^3$). For all simulations of the full Eq. (7.1) random initial conditions are used. Other parameters are chosen as $\varepsilon = 0.05$, $a = 0.5$, $N = 500$, $R_i = 150$, $\phi = \frac{\pi}{2} - 0.1$ and $i = 1, 2, 3$

where $u \equiv u_k^i$ and $v \equiv v_k^i$. Similar to Brandstetter [17], we employ an analytic approximation for the period of the oscillation defined by Eq. (7.8). We consider slow motion on the falling branches of the u-nullcline given by $(1 + \sigma_i + \sigma_{ij})v = u - \frac{u}{3}$ and hence $(1 + \sigma_i + \sigma_{ij})\dot{v} = \dot{u}(1 - u^2)$, which gives:

$$\dot{u} = \frac{(1 + \sigma_i + \sigma_{ij})^2 u + (1 + \sigma_i + \sigma_{ij})a}{1 - u^2}. \tag{7.9}$$

It is possible to integrate this equation analytically from $\pm u_+$ to $\pm u_-$, which are approximately the limits of the slow parts of the u-nullcline, given by $u_+ = 2$ and $u_- = 1$. With this we obtain a rough approximation of the intrinsic period $T(\sigma_{ij})$ of the coupled system, neglecting the fast parts of the trajectory $u(t)$:

$$T(\sigma_{ij}) \propto (1 + \sigma_i + \sigma_{ij})^{-2} \times$$

$$\times \left[u_+^2 - u_-^2 + \left(1 - \left(\frac{a}{1 + \sigma_i + \sigma_{ij}} \right)^2 \right) \ln \frac{a^2 - (1 + \sigma_i + \sigma_{ij})^2 u_-^2}{a^2 - (1 + \sigma_i + \sigma_{ij})^2 u_+^2} \right].$$

$$(7.10)$$

The period T decreases with increasing σ_{ij}. Therefore, due to the resonance condition of τ with respect to the intrinsic period T, the synchronization tongues in Fig. 7.4 are shifted to the left with increasing coupling strength σ_{ij}.

7.2.5 Interplay of Inter- and Intra-Layer Delay

Time delay in complex networks with multiple interacting layers gives rise to special dynamics. In Sect. 7.2.1, we have studied the scenarios of time delay induced patterns in a three-layer network, focusing solely on an inter-layer delay. This kind of delay has been shown to be a powerful tool for controlling various partial synchronization patterns in the three-layer network. Therefore, by multiplexing and introducing inter-layer time delays it is possible to destroy or induce chimera states. In the present Section, we demonstrate that the variation of the delay time in both the intra- and inter-layer links can lead to various dynamical states and allows for control of spatio-temporal patterns. Moreover, in the interplay of inter- and intra-layer time delay we detect full and relay inter-layer synchronization. This Section closely follows [58].

In this Section, we study a multiplex network with three layers (triplex) as shown in Fig. 7.8. In comparison to Sect. 7.2.1, we extend Eq. (7.6) by an intra-layer delay τ_i. By doing so, we will indicate the inter-layer delay as τ_{ij} in this Section:

$$\dot{\mathbf{x}}_k^i(t) = \underbrace{\mathbf{F}(\mathbf{x}_k^i(t))}_{\text{local dynamics}} + \underbrace{\frac{\sigma_i}{2R_i} \sum_{l=k-R_i}^{k+R_i} \mathbf{H}[\mathbf{x}_l^i(t - \tau_i) - \mathbf{x}_k^i(t)]}_{\text{intra-layer coupling}}$$

$$+ \underbrace{\sum_{j=1}^{3} \sigma_{ij} \mathbf{H}[\mathbf{x}_k^j(t - \tau_{ij}) - \mathbf{x}_k^i(t)]}_{\text{inter-layer coupling}},$$

$$(7.11)$$

where all variables are defined as in Eq. (7.1).

To provide an overview of the patterns observed in the network, we calculate the map of regimes in the parameter plane of intra-layer delay time τ_i and inter-layer delay time τ_{ij} (Fig. 7.9). The dominating region is the one corresponding to coherent states (blue region in Fig. 7.9). On the one hand, we detect the in-phase synchronization regime (see Fig. 7.10c), on the other hand, we also observe a region of coherent traveling waves (see Fig. 7.11c). By varying the delay times we can not only switch between these states, but also adjust the speed of traveling waves. In addition, we can observe "salt and pepper" states (green region in Fig. 7.9), where all nodes oscillate with the same phase velocity but they are distributed between states with phase-lag π incoherently [9] (see Figs. 7.10b and 7.11b). The reason for this are strong variations on very short length scales, so that the dynamical patterns have arbitrarily short wavelengths, as we have observed in single-layer systems in Sect. 5.2.2. Besides these two patterns characterized by the same mean phase velocity for all the nodes in the network, we also observe chimera states (red region in Fig. 7.9), where the oscillators within each layer show a characteristic arc-shaped mean phase velocity profile as shown in Figs. 7.10a and 7.11a (right column).

Because of the delay, the system often exhibits traveling structures. Therefore, the classical arc-shaped profile is transformed into a wider, pyramidal or conical one as in Sect. 5.2 (see right column of Figs. 7.10a and 7.11a). We distinguish between the different regions on the one hand, by analyzing the mean phase velocity and a snapshot of variables u_k, on the other hand, by means of the Laplacian distance measure (similar to the spatial correlation coefficient g_0 in Sect. 6.2.1). The Laplacian distance measure identifies strong local curvature in an otherwise smooth spatial profile of $\mathbf{x}_k$ by calculating the discrete Laplacian $\|(\mathbf{x}_{k+1} - \mathbf{x}_k) - (\mathbf{x}_k - \mathbf{x}_{k-1})\|$ for each k. In case of coherent dynamics we obtain low values (≈ 0) for all k, in case of "salt and pepper" states we get high values (> 2). Chimera states show low and high values for the coherent and incoherent domains, respectively. By taking the mean over all k, we can distinguish between coherent, chimera, and "salt and pepper" states.

In many systems with time delays resonance effects can be expected if the delay time is an integer or half-integer multiples of the period in the uncoupled system [28, 34, 78]. Regarding the inter-layer delay time τ_{ij}, we can observe this effect for half-integer multiples of the period $T \approx 2.3$ of a uncoupled FitzHugh-Nagumo

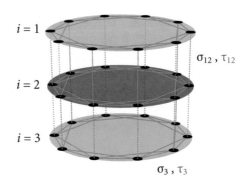

Fig. 7.8 Triplex network with ordinal coupling as in Fig. 7.1: In contrast, the intra-layer coupling is characterized by the strength σ_i and time delay τ_i, and the inter-layer coupling is characterized by the strength σ_{ij} and time delay τ_{ij}. For example, in layer 3 the intra-layer coupling strength is given by σ_3 and the intra-layer time delay is τ_3. Similarly, the inter-layer coupling strength between layer 1 and 2 is given by σ_{12} and the inter-layer time delay by τ_{12}. Figure modified from [58]

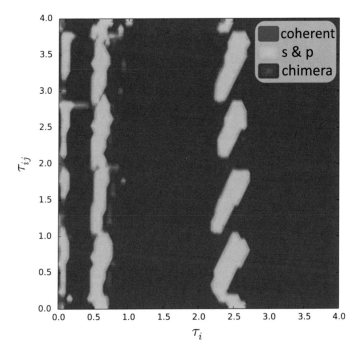

Fig. 7.9 Dynamical regimes in the parameter plane of intra-layer coupling delay $\tau_i \equiv \tau_1 = \tau_2 = \tau_3$ and inter-layer coupling delay $\tau_{ij} \equiv \tau_{12} = \tau_{23}$: "salt and pepper" states (green islands) occur in the region of coherent states (blue region) as traveling waves, cluster or synchronized states. At the border between these two regimes chimera states can be found (red color). The distinction between the different regions are detected on the one hand, by analyzing the mean phase velocity and a snapshot of variables u_k, on the other hand, by means of the Laplacian distance measure (see Sect. 6.2.1). The boundary of these regions are fitted linearly after the (τ_i, τ_{ij}) plane has been sampled in steps $\Delta\tau_i = 0.05$ and $\Delta\tau_{ij} = 0.1$. For all simulations of Eq. (7.11) random initial conditions are taken. Parameters are chosen as $\varepsilon = 0.05$, $a = 0.5$, $\sigma_i = 0.2$, $\sigma_{ij} = 0.05$, $N = 500$, $R_i = 170$, $\phi = \frac{\pi}{2} - 0.1$, and $i, j = 1, 2, 3$. Figure from [58]

oscillator: In Fig. 7.9, the green island are repeating for $\Delta\tau_{ij} \approx T/2$ at a fixed $\tau_i \approx 2.5$. Concerning the intra-layer delay time τ_i, we find a resonance effect in the case of integer multiples of the delay. For greater values of the delay times τ_i and τ_{ij} the dynamical regions are becoming curved (see green islands in Fig. 7.9 at $\tau_i \approx 2.5$). This can be explained by the fact that branches of periodic solutions, which are reappearing for integer multiples of the intrinsic period, are becoming stretched with increasing delay time as shown in Sect. 5.2.3. In comparison to the almost vertical shape at $\tau_i \approx 0.5$, the green islands at $\tau_i \approx 2.5$ are rotated clockwise by approximate $\pi/8$. The consequence is an overlapping of the delay islands for small intra-layer delay time τ_i, whereas for larger delays the islands become separated. Therefore, for $\tau_i = 0$, we can observe chimera states almost independently of the inter-layer time delay τ_{ij}. This scenario has been discussed in Sect. 7.2.1 (see Fig. 7.4).

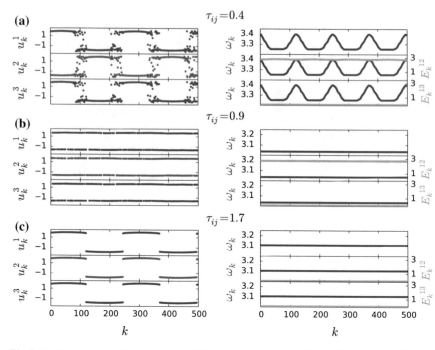

Fig. 7.10 Dynamics in all three layers for different values of the inter-layer delay time τ_{ij}: **a** Chimera state for $\tau_{ij} = 0.4$, **b** "salt and pepper" state for $\tau_{ij} = 0.9$, **c** coherent state (cluster state) for $\tau_{ij} = 1.7$ The intra-layer delay time is fixed at $\tau_i = 0.8$. The left column displays snapshots of variables u_k^i for the layers $i = 1, 2, 3$, while the right column illustrates the mean phase velocity profile ω_k^i (dark blue) for the individual layers and the local inter-layer synchronization error E_k^{ij} (light yellow) calculated from Eq. (7.7). Other parameters as in Fig. 7.9. Figure is taken from [58]

In Figs. 7.10 and 7.11 (right column), E_k^{ij} is plotted (light yellow) together with the mean phase velocity (blue): Depending on the delay times τ_i and τ_{ij}, we can find full inter-layer synchronization (see Fig. 7.10c) as well as relay inter-layer synchronization (see Figs. 7.10a, b and 7.11a, c). These synchronization scenarios can be found for both coherent and incoherent dynamics. An additional effect is the partial relay synchronization scenario in Fig. 7.11b: In all three layers "salt and pepper" dynamics can be observed. The nodes in the outer layers are almost all synchronized but a small part of them destroys the relay synchronization. On the other hand a few oscillators in the relay layer are synchronized with the outer layers.

To sum up, in the parameter plane of the intra-layer τ_i and the inter-layer τ_{ij} time delay, we have determined the regions where chimera patterns occur, alternating with regimes of coherent states. A proper choice of time delays allows to achieve the desired state of the network: chimera state or coherent pattern, full or relay inter-layer synchronization. Combining the delayed interactions with the multiplex framework considered in this Section can provide additional insight into the formation of the complex spatio-temporal patterns in real-world systems. Specifically, this is of inter-

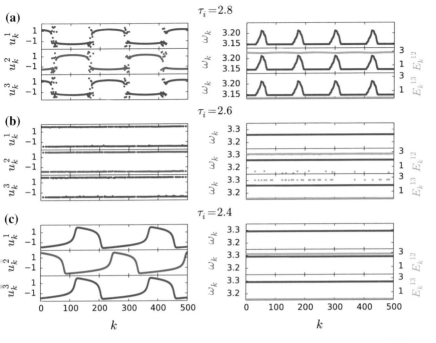

Fig. 7.11 Dynamics in all three layers for different values of the intra-layer delay time τ_i: **a** Chimera state for $\tau_i = 2.8$, **b** "salt and pepper" state for $\tau_i = 2.6$, **c** coherent state (traveling wave) for $\tau_i = 2.4$. The inter-layer delay time is fixed at $\tau_{ij} = 2.6$. The left column displays snapshots of variables u_k^i for the layers $i = 1, 2, 3$, while the right column illustrates the mean phase velocity profile ω_k (dark blue) for the individual layers and the local inter-layer synchronization error E_k^{ij} (light yellow) calculated from Eq. (7.7). Other parameters as in Fig. 7.9. Figure is taken from [58]

est in the context of brain networks where electroencephalography (EEG) patterns are recently reported to display chimera-like behavior at the onset of a seizure [4, 22, 53]. Inducing the chimera states by tuning the inter- and intra-layer delay values provides us with a powerful tool to control chimera states.

7.2.6 Detrending Method

Generally, introducing a time delay τ in the intra- or inter-layer coupling term often leads to traveling patterns as shown, e.g., in Fig. 5.8 of Chap. 5. Additionally, chimera states are associated with complex dynamics sensitively depending on initial conditions and it is difficult to obtain general dependencies of the velocity of the traveling pattern on the delay or the coupling strength. For our multiplex network in case of delayed coupling, we observe the same effect (see Fig. 7.12c).

Consequently, it is not possible to extract any information from measures calculated over a long time, e.g., the mean-phase velocity profile and local inter-layer synchronization error E_k^{ij} (see Fig. 7.12a). By detrending the data we can avoid this problem: After each time-step in the numerical simulation we re-index the nodes k in such a way that $k' = (k + c)|_N$, where $|$ stands for the modulo operator and c is given by the center of the largest domain of the ring where for all k's of that domain $\|x_k(t) - x_{k+1}(t)\| < \theta$. As the threshold parameter θ we choose $\theta = 0.25$. The effect of that method is shown in Fig. 7.12: Without detrending (Fig. 7.12a) no information can be achieved from the mean phase velocity profile, the local or global inter-layer synchronization error (see Eq. 7.7); applying detrending leads to well pronounced arc-shaped profiles of both measures (Fig. 7.12b). The mechanism of the detrending method can be interpreted as the comoving frame of the center of the coherent domain.

To be more specific, in Fig. 7.12, E_k^{13} is plotted (light yellow) together with the mean phase velocity ω_k (dark blue) for a typical chimera state. Since the chimera patterns are traveling along the ring, averaging over a time window does not lead to a clear chimera-type profile (Fig. 7.12a). However, after re-indexing the nodes

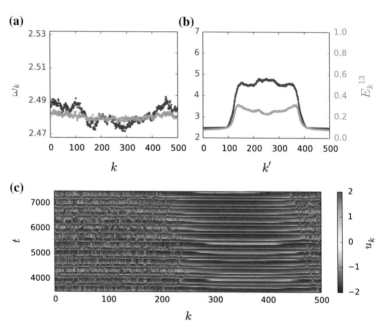

Fig. 7.12 Detrending of simulation data: By re-indexing the nodes for each numerical step we can detect chimera states which shift temporally along the ring, for instance, induced by delayed coupling. Upper left panel **a** shows raw simulation data for the mean phase velocity ω_k (dark blue) and local inter-layer synchronization error E_k^{13} (light yellow) calculated from Eq. (7.7), whereas the upper right panel **b** shows the same data after detrending, i.e. re-indexing. A typical arc-shape profile is recognizable now. The lower panel **c** shows a space-time plot of the variable u_k for traveling chimera inside a layer. Simulations are performed for $\sigma_i = 0.2$, $\sigma_{ij} = 0.01$, $R_1 = R_3 = 150$, $R_2 = 110$, and $\tau = 0.7$. Other parameters as in Fig. 7.5. Figure modified from [57]. © 2018 by the American Physical Society (APS)

after each time-step such that the coherent domain remains at a fixed spatial position ("detrending") we can see the arc-shaped profiles for both mean phase velocity ω_k and local inter-layer synchronization error E_k^{13} (Fig. 7.12b). Chimera states can thus also be detected in traveling structures. It has to be mentioned that the value of the mean phase velocity ω_k is increased in the incoherent part. This increment depends on the simulation time and exists due to the fact that the largest coherent domain, as defined above, has not a constant width but its width varies in time chaotically. Therefore, the center of the largest coherent domain follows that chaotic movement and has an effect on numerical counting methods.

7.3 Partial Relay Synchronization

In the present Section, we extend the notion of relay synchronization from completely synchronized states to partial synchronization patterns in the individual layers. This section closely follows [56, 57] [for [57]: © 2018 by the American Physical Society (APS)]. We analyze special regimes where only the coherent domains of chimeras are synchronized, and the incoherent domains remain desynchronized, as well as transitions between different synchronization scenarios. Varying the coupling strength and time delay in the inter-layer connections, we observe so-called partial relay synchronization between chimera states in the outer network layers. In Sect. 7.3.1, we will characterize these novel states. To detect partial relay synchronization, we take advantage of the local inter-layer synchronization error, a useful measure, which we have introduced in Sect. 7.2.2. Changing the coupling range in Sect. 7.3.2, i.e., the topology, of the relay layer allows to establish its effect on the remote synchronization in the outer layers. The phenomenon of partial relay synchronization will be also investigated in case of mismatched excitation parameters in Sect. 7.3.3. In Sect. 7.3.4, we will present a detailed comparison with experimental results.

7.3.1 Double Chimera States

In Sect. 7.2.2, partial relay synchronization has been briefly touched as we introduced the local inter-layer synchronization error E_k^{ij}. Dynamics inside the (red dotted) region of desynchronization in Fig. 7.4 exhibit a remarkable structure. In Fig. 7.5c, we can see the arc-shaped profiles for both mean phase velocity ω_k and local inter-layer synchronization error E_k^{13}. This means that the coherent parts of the chimera states are synchronized between the outer layer, whereas the incoherent parts are not. This kind of synchronization may be called *partial relay inter-layer synchronization* or *double chimera*, since it denotes coherence-incoherence behavior within the layers and between the layers. It cannot be detected by the global inter-layer synchronization error $E^{ij} \neq 0$, but the node-dependent local measure E_k^{ij} gives us the possibility to distinguish this type of synchronization. In Fig. 7.5 (right column), E_k^{ij} is plotted (light

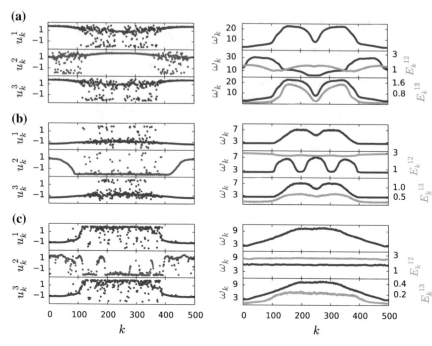

Fig. 7.13 Partial relay inter-layer synchronization between the outer layers: snapshots of variable u_k^i (left column) for layers $i = 1, 2, 3$ (relay layer: light red, outer layers: dark blue), and mean phase velocity profiles ω_k (dark blue) and inter-layer synchronization error E_k^{ij} (light yellow) in the right column. **a** $R_1 = R_2 = R_3 = 170$, $\tau = 1.3$, $\sigma_{ij} = 0.025$; **b** $R_1 = R_3 = 150$, $R_2 = 130$, $\tau = 0.4$, $\sigma_{ij} = 0.015$; **c** $R_1 = R_3 = 150$, $R_2 = 10$, $\tau = 0.8$, $\sigma_{ij} = 0.01$. Other parameters as in Fig. 7.4. Figure from [57]. © 2018 by the American Physical Society (APS)

yellow) together with the mean phase velocity profile ω_k (dark blue) for a typical chimera state.

In our simulations, we observe different intriguing types of partial relay inter layer synchronization, for instance, Fig. 7.13a depicts an example ($\tau = 1.3$) where the relay layer exhibits anti-synchronization of chimera patterns: the coherent domain of the relay layer (red, middle panel) spatially coincides with the incoherent domain of the outer layers. We will focus on these "anti"-synchronized patterns in more detail in Sect. 7.5.

7.3.2   Robustness Against Mismatched Coupling Radii

In this Section, we examine synchronization scenarios in the case of mismatched coupling radii $R_{1/3} \neq R_2$. In particular, we will focus on partial relay synchronization. To investigate this recently discovered phenomenon [56, 57] in more detail we will focus solely on the inter-layer delay τ as the control parameter (see Sect. 7.2.1). In contrast, the inter-layer delay has been denoted by τ_{ij} in Sect. 7.2.5.

We introduce a mismatch of $R_{1/3} - R_2 = 20$ in our system. Figure 7.13b shows partial relay inter-layer synchronization for the case of small mismatch of the coupling range in the relay and outer layers ($R_1 = R_3 = 150, R_2 = 130$). The middle layer exhibits a chimera state with three incoherent domains, in contrast to two in the outer layers. Moreover, the coherent domains in the relay layer and the outer layers are in anti-phase. The local synchronization error E_k^{13} between the two outer layers is nonzero in the incoherent domains and vanishes for the coherent domains, as a signature of partial relay synchronization.

Furthermore, we investigate the interplay of mismatched coupling radii and inter-layer delay τ. Depending on which values of τ we choose in Eq. (7.6), we can observe either full or relay inter-layer synchronization. In Fig. 7.14, we can identify full in-phase synchronization of all three layers for values of τ close to integer multiples of the period of the uncoupled system $T \approx 2.3$, and relay inter-layer synchronization with anti-phase synchronization between the outer layers and the relay layer for half-integer multiples.

In Fig. 7.15, we have also a smaller coupling range in the relay layer ($R_2 = 110$) compared to the outer layers ($R_1 = R_3 = 150$). This Fig. 7.15 shows two examples: for $\tau = 0.3$ we can observe chimera states in each layer (snapshots in Fig. 7.15a and mean phase velocity profiles ω_k in Fig. 7.15b). In the outer layers the coherent parts are synchronized, whereas the incoherent parts are not (light yellow curve z_k^{13}) (see bottom panel of Fig. 7.15b). Due to the larger mismatch of coupling radii

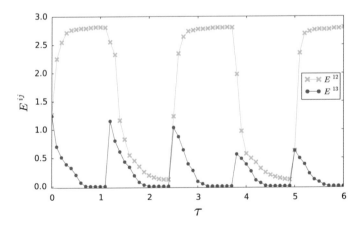

Fig. 7.14 Influence of delay time τ on the synchronizability in a multiplex network: Global inter-layer synchronization error E^{ij} versus time delay τ between the layers. Between the first and third layer (dark blue filled circles) we can observe synchronization for values of τ close to half-integer multiples of the period of the uncoupled system $T = 2.3$ whereas the relay layer is in anti-phase for half-integer multiples and in-phase for integer multiples (light yellow crosses). Parameters: $\sigma_i = 0.2, \sigma_{ij} = 0.02, R_1 = R_3 = 150$, and $R_2 = 110$. Other parameters as in Fig. 7.2. Figure from [56]

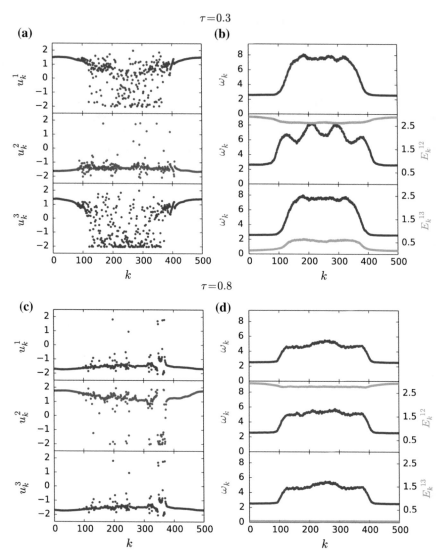

Fig. 7.15 Dynamics inside the three layers for two values of the delay time τ: **a, b** "partial" relay inter-layer synchronization between the outer layers for $\tau = 0.3$. (**c, d**) "full" relay inter-layer synchronization for $\tau = 0.8$. The left column **a, c** shows snapshots of variables u_k^i for all three layers $i = 1, 2, 3$ (relay layer: light red, outer layers: dark blue), whereas the right column **b, d** shows the mean phase velocity profile ω_k (dark blue) for each layer and the local inter-layer synchronization error E_k^{ij} (light yellow). Other parameters as in Fig. 7.14. Figure modified from [56]

$R_{1/3} - R_2 = 40$ the middle layer exhibits a 4-chimera state, where the coherent domain is in anti-phase to the coherent domains of the outer layers (see middle panel of Fig. 7.15b). This kind of synchronization may be called a "partial" relay inter-layer synchronization. By increasing the delay to $\tau = 0.8$ in Fig. 7.15c and d, we can achieve "full" relay inter-layer synchronization between the outer layers ($E_k^{13} = 0$). Both coherent and incoherent domains are now synchronized. In the middle layer the 4-chimera vanishes and a chimera state similar to the outer layers appears inspite of the different coupling range $R_2 \neq R_{1/3}$, and the coherent domain is still in anti-phase to the coherent domains of the outer layers.

Moreover, for large mismatch of the coupling ranges in the relay and outer layers $R_{1/3} - R_2 = 140$, the relay layer is characterized by chaotic dynamics. In Figs. 7.13c and 7.16, we choose a small coupling range $R_2 = 10$ in the relay layer, whereas $R_1 = R_3 = 150$. Due to that small coupling range, the strongly chaotic dynamics of the relay naturally affects the chimera states in the outer layers, so that their mean phase velocity profiles (dark blue) are smeared out despite of detrending. Nevertheless, the coherent domains of the chimera states are synchronized between the outer layers, whereas the incoherent parts are not, as shown in the snapshots and the plot of E_k^{13}. Thus, the relay synchronization mechanism turns out to be resilient against changes of the relay layer topology.

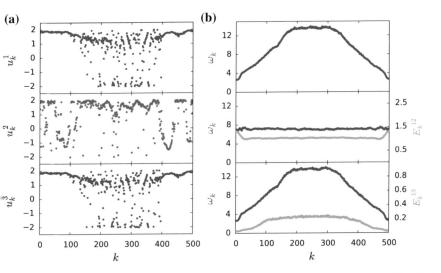

Fig. 7.16 Partial relay inter-layer synchronization with chaotic dynamics in the relay layer: The left column **a** shows snapshots of u_k^i for all three layers $i = 1, 2, 3$ (relay layer: light red, outer layers: dark blue), whereas the right column **b** represents the mean phase velocity profile ω_k (dark blue) for each layer and the inter-layer synchronization error E_k^{ij} (light yellow). The coherent domains of the chimera state are synchronized between the outer layers, whereas the incoherent parts are not. Due to the small coupling range $R_2 = 10$ in the relay layer, chaotic dynamic occurs. Parameters: $\tau_i = 0.2, \sigma_{ij} = 0.01, R_1 = R_3 = 150$, and $\tau = 0.7$. Other parameters as in Fig. 7.2. Figure modified from [56]

To sum up, we have varied the topology of the relay layer compared to the outer layers by changing its coupling range, but the phenomenon of partial relay synchronization turns out to be robust.

7.3.3   Robustness Against Mismatched Excitation Parameters

In general, there exist several parameters which can render the relay layer different from the outer layers. So far, we have focused on the topology R_i to make the relay layer different from the outer layers. Another possibility is to vary the dynamics of the nodes. Preliminary studies show that the phenomenon of partial relay synchronization holds also for a mismatch of the excitation parameter a between the layers. For this purpose we modify Eq. (7.6) by introducing a layer-dependent a_i. The dynamics of each individual oscillator is governed now by:

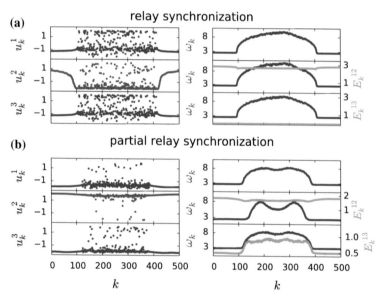

Fig. 7.17 Dynamics of the three layers for mismatched excitation parameters $a_1 = a_3 = 0.5$ and $a_2 = 0.6$: **a** Relay inter-layer synchronization for $\tau = 5.5$. **b** Partial relay inter-layer synchronization between the outer layers for $\tau = 0.4$. The left column shows snapshots of variables u_k^i for all three layers $i = 1, 2, 3$ (relay layer: light red, outer layers: dark blue), whereas the right column shows the corresponding mean phase velocity profiles ω_k (dark blue) for each layer and local inter-layer synchronization error E_k^{ij} (light yellow). Inter-layer coupling is given by $\sigma_{ij} = 0.025$, intra-layer coupling is given by $\sigma_i = 0.2$, other parameters as in Fig. 7.4. Figure from [57] © 2018 by the American Physical Society (APS)

$$\mathbf{F(x)} = \begin{pmatrix} \varepsilon^{-1}(u - \frac{u^3}{3} - v) \\ u + a_i \end{pmatrix},$$ (7.12)

where each layer i has a specific threshold parameter a_i. In Fig. 7.17, we show that the phenomenon of (partial) relay synchronization can also be observed in the case where there is a mismatch between the excitation parameters in the outer layers and the relay layer ($a_1 = a_3 \neq a_2$).

7.3.4 Comparison with Experimental Results

Recently it has been asserted that several regions in the brain, e.g., the thalamus and hippocampus, act as a relay element between two distant regions in the brain [37]. In [32] the authors suggest that synchronization between two neocortical regions in the brain is likely to be mediated by the hippocampus as a relay layer. They report that two distant neuronal populations, namely the frontal (F) and visual (V) cortex, are able to synchronize at almost zero time-lag if a third element, the hippocampus (H), acts as a relay between them (see Fig. 7.18a). This relay symmetrically redistributes its incoming signals between the two other regions. Experiments suggest that zero-lag neuronal synchrony occurs in the brain in the presence of large axonal conduction delays [25]. Moreover, it has been shown in experiments with mice that the relay element is synchronized with a constant phase-lag with respect to the two cortical regions (see Fig. 7.18b). To simulate these findings the authors in [32] suggest a network model, where the two different regions F, V of size $N = 500$ are not only connected via the relay element H, but are also mutually coupled (see right column of Fig. 7.18a).

In our model of a triplex network (same size as [32]) we show that by introducing delay in the inter-layer connections one is able to achieve relay synchronization, where no direct connection between the outer layers is necessary, which is more in line with the real structure of the human brain. Furthermore, in our simulations (e.g., Fig. 7.17, or Figs. 7.5b, c and 7.13) the relay layer is synchronized with a constant phase lag similar to the experiments in [32]. Finally, with our scenario of double chimera (partial relay inter-layer synchronization) we are able to propose an explanation of the imperfect synchronization shown in Fig. 7.18b, c: For special delay times just a part of the outer layers is synchronized, whereas some parts stay desynchronized.

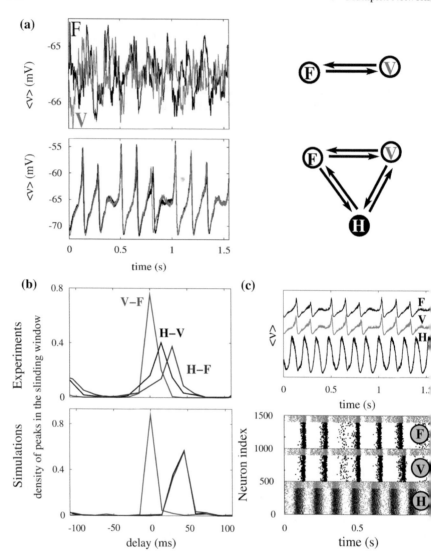

Fig. 7.18 Comparison with experimental results [32]: **a** Measured time series (left column) of average voltage v in mV of two cortical regions, frontal (F, black) and visual (V, light orange). Upper panel shows desynchronization and corresponds to the case of a 2-layer network (see right column), whereas the lower panel shows zero-lag synchronization and corresponds to a 3-layer network, where the two cortical regions are additionally connected with the hippocampal relay (H). **b** Spatio-temporal synchronization obtained from the experimental (upper panel) and numerical data (lower panel): Plotted is the density of spikes in the sliding window (300 ms) of filtered time series cross-correlation between the three layers F, V, H. Experimental data correspond to a mouse. **c** Modeling of neuronal dynamics: Lower panel: space-time plot of 500 neurons in each layer. Upper panel: ensemble average voltage of the three lays versus time in s. All panels after [32]

7.4 Relay Synchronization in Networks of Time-Discrete Maps

In the last Sections, we have dealt with time-continuous oscillators. Since their initial discovery, chimera states have evoked great interest [15, 16, 24, 45, 47, 48, 51, 66, 77] and have been found in numerous models, among them time-discrete maps with chaotic and periodic dynamics [13, 14, 18, 19, 49, 62, 65, 67, 68, 72]. In this Section, we study relay synchronization in three-layer multiplex networks of time-discrete logistic maps, where the individual uncoupled maps are characterized by chaotic dynamics, and inside each layer we have a nonlocally coupled ring. We demonstrate the synchronization of chimera states in the two outer layers of our network, which interact via the intermediate layer. Moreover, we study the transition from phase to amplitude chimeras with increasing inter-layer coupling strength and provide analytical conditions for phase chimeras in terms of the network coupling strength. This Section is closely related to [75, 76].

7.4.1 Time-Discrete Logistic Maps

In Sect. 2.1.2, we have introduced the logistic map as a prominent example of a time-discrete dynamical system, where a simple nonlinearity in the equation yields complex and chaotic behavior. The system unter investigation is a multiplex network consisting of three layers, as illustrated in Fig. 7.1. Each layer is represented by a ring of non-locally coupled time-discrete logistic maps:

$$
\begin{aligned}
z_k^i(t+1) = \underbrace{f(z_k^i(t))}_{\text{local dynamics}} &+ \underbrace{\frac{\sigma_i}{2R_i} \sum_{l=k-R_i}^{k+R_i} \left[f(z_l^i(t)) - f(z_k^i(t)) \right]}_{\text{intra-layer coupling}} \\
&+ \underbrace{\sum_{j=1}^{3} \sigma_{ij} \left[f(z_k^j(t)) - f(z_k^i(t)) \right]}_{\text{inter-layer coupling}},
\end{aligned}
\tag{7.13}
$$

where z_k^i are real dynamical variables, with node index $k = 1, \ldots, N$ and layer index $i = 1, 2, 3$; all indices k, l are modulo N. The discrete time is denoted by t; $f(z)$ is a one-dimensional logistic map in Eq. (2.17), where we fix the bifurcation parameter $a = 3.8$, corresponding to chaotic dynamics of the individual uncoupled unit. Earlier studies have shown that other choices of the parameter a result in qualitatively similar findings [50]. R_i is the non-local intra-layer coupling range, associated with the dimensionless coupling radius $r_i = \frac{R_i}{N}$, and σ_i is the intra-layer coupling strength.

For an ordinal inter-layer coupling with constant row sum we choose the inter-layer coupling matrix as in Eq. (7.3).

In each individual layer, it is possible to observe chimera patterns depending on the range of nonlocal coupling and the strength of the couplings [49]. Note that chimera states in networks of time-discrete coupled maps differ from the chimera states in networks of time-continuous oscillators [52]. Chimera patterns, consisting of coherent and incoherent domains, emerge as a result of the break-up of the smooth wavelike profiles [35, 49, 50], with an even number of coherent domains, and successive coherent domains appear in anti-phase to one another, while the incoherent domains are characterized by spatial chaos. Although the dynamics of the uncoupled map is chaotic, chimera states perform periodic dynamics in time, and this period depends on the system parameters [49]. Therefore, the mean frequency which is used as main criterion for chimeras in time-continuous oscillatory systems is not applicable in the case of time-discrete maps. To analyze the synchronization of such complex patterns between the layers of our three-layer network, we will fall back upon the global and local inter-layer synchronization error in Eqs. (7.5) and (7.7). Due to the time-discrete systems, we have to modify the measures in an appropriate way:

$$E^{ij} = \lim_{T \to \infty} \frac{1}{NT} \sum_{t=0}^{T} \sum_{k=1}^{N} \left\| z_k^j(t) - z_k^i(t) \right\|, \qquad (7.14)$$

where $\|\cdot\|$ is the Euclidean norm and T is the number of time-steps. To characterize the synchronization of chimera patterns between the layers in more detail, we define the local synchronization error in similar way:

$$E_k^{ij} = \lim_{T \to \infty} \frac{1}{T} \sum_{t=0}^{T} \left\| z_k^j(t) - z_k^i(t) \right\|. \qquad (7.15)$$

Figure 7.19 presents simulations of the triplex network in the regime where the individual uncoupled layers exhibit chimera states. The plot shows the global synchronization error E^{ij} (Eq. 7.14) versus the inter-layer coupling strength $\sigma_{ij} = \sigma_{12} = \sigma_{32}$. The green dots represent the global synchronization error E^{13} between the first and the third layer. The red stars represent the global synchronization error E^{12} between the first and the second layer. Specially prepared initial conditions as in [35, 49] have been used to induce chimera states faster. The population of each layer is divided into four equal parts where two of them (not neighboring) are initialized with uniform initial conditions, and the other two parts are initialized with a randomly alternating sequence of two initial values, one low (0.4455) and one high (0.8436). In order to avoid immediate inter-layer synchronization, we have shifted the four parts in each layer with respect to their position in the other layers. Hence, every ring in the three-layer network has its own initial conditions.

The synchronization error between the first and third layer stays below the error between the first and second layer. This is an indicator of relay synchronization where the second layer plays the role of a relay layer. The overall synchronization of the considered 3-layer network tends to increase with increasing inter-layer coupling strength. At a certain value of σ_{ij} the synchronization error drops sharply within a short range of σ_{ij} to zero where E^{13} reaches zero before E^{12}.

Exemplary snapshots of the dynamics for selected inter-layer coupling strength σ_{ij} from regions of interest of Fig. 7.19 are shown in Fig. 7.20: Figure 7.20a–c depict the dynamical variable z_k^1, z_k^2, z_k^3 of the first, second and third layer, respectively. Figure 7.20d, e represent the local synchronization error between the first and second layer E_k^{12} and between the first and third layer E_k^{13}, respectively, for each node as introduced in Eq. (7.15). This is an adequate space-resolved measure for the synchronization of chimera patterns.

The first column A in Fig. 7.20 corresponds to weak inter-layer coupling strength $\sigma_{ij} = 0.015$ and to a large value of E^{13} marked by a black square in Fig. 7.19. The low value of σ_{ij} has almost no effect on the triplex network since no synchronization can be observed in the snapshots. Also the local synchronization errors E_k^{12}, E_k^{13} in Fig. 7.20d, e show high values.

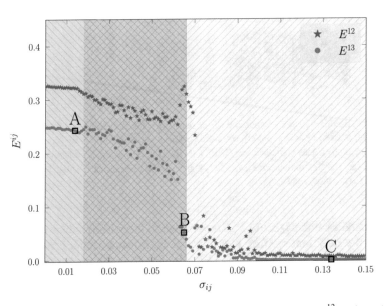

Fig. 7.19 Global synchronization error between the first and second layer E^{12} (red stars) and between the first and third layer E^{13} (blue dots), respectively, versus σ_{ij}. For small inter-layer coupling strength the system is completely desynchronized, and with increasing σ_{ij} the overall synchronization increases. The regimes of phase chimeras and amplitude chimeras are shown by diagonal stripes going from the upper left (blue) and right corner (gray shading), respectively. Note that there is an overlap between the two regimes. The black squares A (left), B (middle) and C (right) mark three values of σ_{ij} which are further analyzed in Fig. 7.20. Other parameters: $\sigma_i = 0.22$, $\sigma = 0.28$, $N_i = 1000$, $a = 3.8$. All simulations are run for 5000 time steps and the synchronization error is averaged over the last 50 steps to avoid transient effects. Figure modified from [76]

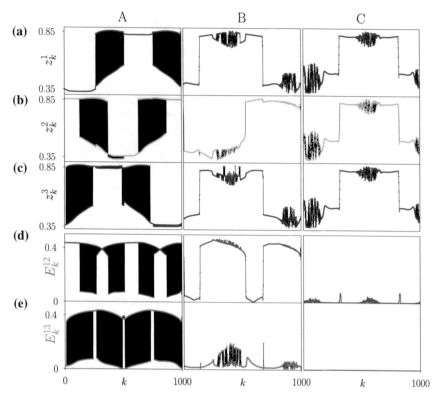

Fig. 7.20 Transition to relay synchronization for three selected values of inter-layer coupling strength σ_{ij}, marked by A (left panel), B (middle panel) and C (right panel) for $\sigma_{ij} = 0.015, 0.066$ 0.135 respectively (black squares in Fig. 7.19). **a–c**: Snapshots of z_k^1, z_k^2, z_k^3, **d** local synchronization error between the first and second layer E_k^{12}, **e** E_k^{13} versus node index i. Other parameters as in Fig. 7.19. Figure modified from [76]

The middle column B in Fig. 7.20 is associated with an intermediate value of $\sigma_{ij} = 0.066$ in the transition zone from high to low global synchronization error E^{13} marked by a black square in Fig. 7.19. In Fig. 7.20, the snapshots of the dynamical variables show slightly different small-amplitude oscillation but spatial synchrony (Fig. 7.20a c). The second layer (Fig. 7.20b) shows desynchronized behavior compared to the two outer layers, i.e., the coherent and incoherent domains are incongruous except for a small overlap (see Fig. 7.20d). The stronger inter-layer coupling leads to adaptation of the location of incoherent and coherent domains on the two rings in the outer layer mediated by the relay layer. Nevertheless, the oscillations of each node are still not perfectly synchronized, which is shown by the global synchronization error E^{13} in panel (Fig. 7.20e): The coherent part of the outer layers are synchronized, whereas the incoherent ones are not. As mentioned before, we call such a scenario *partial relay synchronization*.

The third column C in Fig. 7.20 extracts further information about the behavior of the system for large inter-layer coupling strength $\sigma_{ij} = 0.135$ marked in Fig. 7.19 by a black square. Here, full relay synchronization can be observed. The coherent and incoherent domains coincide in the first and third layer (see Fig. 7.20a, c). Also the second layer (Fig. 7.20b) behaves qualitatively in a similar way, however, comparing the local synchronization error, one can see that E_k^{13} is exactly equal to zero for each node in the layer, whereas the local synchronization error $E_k^{12} \neq 0$ demonstrates that the second layer is not completely synchronized with its outer counterparts. It acts like a relay layer which stabilizes synchrony between the two outer layers, which is a signature of relay synchronization between the outer layers of the network.

7.4.2 Transition from Phase to Amplitude Chimeras

In continuous-time oscillator networks, relating to the amplitudes of the oscillators, two different types of chimera states have been observed and studied. The so-called *phase chimeras* demonstrate coexistence of oscillator groups with synchronized and desynchronized phases. The *amplitude chimeras* are patterns where all oscillators are phase locked, but synchronized and desynchronized domains appear in terms of oscillator amplitudes [10, 11, 80]. In recent works on networks of time-discrete coupled maps, these two notions have been also used to define phase and amplitude chimeras in terms of incoherent domains with large (phase flips) and small variations, respectively. In contrast to the phase chimera states in [49], amplitude chimera states are characterized by completely asynchronous chaotic dynamics in the incoherent domains [13, 14]. In this Chapter, we will use this terminology as well.

The regimes of phase chimeras and amplitude chimeras are shown in Fig. 7.19 by diagonal stripes going from the upper left (blue) and right corner (gray shading), respectively. Note that there is an overlap between the two. Figure 7.20 demonstrates three examples of patterns in each of three layers for small, intermediate and large inter-layer coupling strength, marked as A, B, C in Fig. 7.19. For weak inter-layer coupling strength (A), we observe phase chimera states in each layer of the multiplex network. The first column in Fig. 7.20 shows in black the incoherent domains in all three layers where we find an irregular sequence of values on the upper and lower coherent branch, corresponding to phase flips of π. Pronounced phase chimera states like these cannot be observed for larger values of σ_{ij}.

With increasing inter-layer coupling strength σ_{ij}, amplitude chimera states are born within the upper and lower coherent branch (B). They are characterized by small amplitude variations in the incoherent domains, rather than phase flips of π, and arise since each layer consists of a closed-ring of time-discrete oscillators which are additionally coupled to an extra node from another layer. These amplitude chimeras can be perfectly synchronized in the considered network between the outer layers as indicated by the local synchronization error $E_k^{13} = 0$ in the right column (C) of Fig. 7.20e. On the other hand, $E_k^{12} \neq 0$ proves that the middle layer is not fully synchronized, i.e., relay synchronization arises. The corresponding plots of E^{12} and

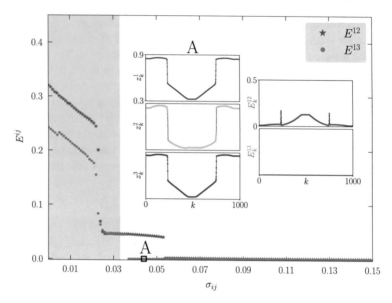

Fig. 7.21 Global synchronization error between the first and second layer E^{12} (red stars) and between the first and third layer E^{13} (blue dots), respectively, versus σ_{ij} for $\sigma_i = 0.32$, $r_i = 0.30$. The inset shows the snapshots and local synchronization errors E_k^{12}, E_k^{13} corresponding to the black square A in the main figure $\sigma_{ij} = 0.045$. Blue color denotes the regime of phase chimeras. Other parameters as in Fig. 7.19. Figure modified from [76]

E^{13} in Fig. 7.20 illustrate that the necessary condition for phase chimeras derived below is violated.

The intra-layer parameters play also an important role in the synchronization scenarios for the whole network. Figure 7.21 shows the dependence of the global synchronization error on the inter-layer coupling strength in the case of larger coupling range $r_i = 0.3$ inside the layers and stronger intra-layer coupling strength $\sigma_i = 0.32$. In this case, only very weak inter-layer coupling results in the observation of phase chimeras in all layers, and its further increase moves the system towards the formation of smooth profiles. However, there is indeed an intermediate region where distinct relay synchronization between the outer layers occurs, bounded by two abrupt discontinuous transitions of E^{13} and E^{12} to zero, and all this occurs for smooth profiles in contrast to Fig. 7.19. The insets in Fig. 7.21 depict the corresponding snapshots and local synchronization errors at point A.

Using a similar argument as in [35, 50, 62] for single-layer networks, we can calculate analytically the critical intra-layer coupling strength σ_c for the onset of phase chimeras in case of a triplex network. The coherent solutions $z_k^i(t)$ approach a smooth profile $z(x^i, t)$ in the continuum limit $N \to \infty$ which leads to one equation per layer, where x^i denotes the space coordinate in layer i. For a clear calculation we denote the intra-layer coupling strength $\sigma_i = \sigma_1 = \sigma_2 = \sigma_3 \equiv \sigma$, the inter-layer coupling strength $\sigma_{ij} = \sigma_{12} = \sigma_{23} \equiv \Phi$, and the coupling range $R_1 = R_2 = R_3 = R$

hence $r_i = \frac{R_i}{N} = r$ in this Section. Thereby, we obtain the following equations for each layer:

$$z^{(1)}(x, t+1) = f\left(z^{(1)}(x, t)\right) + \frac{\sigma}{2r} \int_{x-r}^{x+r} \left[f(z^{(1)}(y, t)) - f(z^{(1)}(x, t))\right] dy$$
$$+ \Phi\left[f(z^{(2)}(x, t)) - f(z^{(1)}(x, t))\right], \tag{7.16}$$

$$z^{(2)}(x, t+1) = f\left(z^{(2)}(x, t)\right) + \frac{\sigma}{2r} \int_{x-r}^{x+r} \left[f(z^{(2)}(y, t)) - f(z^{(2)}(x, t))\right] dy$$
$$+ \frac{\Phi}{2}\left[f(z^{(1)}(x, t)) - f(z^{(2)}(x, t))\right]$$
$$+ \frac{\Phi}{2}\left[f(z^{(3)}(x, t)) - f(z^{(2)}(x, t))\right], \tag{7.17}$$

$$z^{(3)}(x, t+1) = f\left(z^{(3)}(x, t)\right) + \frac{\sigma}{2r} \int_{x-r}^{x+r} \left[f(z^{(3)}(y, t)) - f(z^{(3)}(x, t))\right] dy$$
$$+ \Phi\left[f(z^{(2)}(x, t)) - f(z^{(3)}(x, t))\right]. \tag{7.18}$$

To derive a relation for the critical intra-layer coupling strength σ_c we can conduct the following steps. First, we transform the equation for layer 1 into:

$$z^{(1)}(x, t+1) = (1 - \sigma - \Phi)f\left(z^{(1)}(x, t)\right)$$
$$+ \frac{\sigma}{2r} \int_{x-r}^{x+r} f\left(z^{(1)}(y, t)\right) dy + \Phi f\left(z^{(2)}(x, t)\right). \tag{7.19}$$

Consider a solution of Eq. (7.19) in the form of a smooth wave profile with wave number $k_w = 1$, i.e., with wavelength λ equal to the length of the ring L, and period-2 dynamics for each layer. We can reduce the dynamics to even and odd time steps $z_0(x)$ and $z_1(x)$, respectively. This yields the following spatial derivatives for $j = 0, 1$ where we distinguish between two cases, where in the first case the two layers 1 and 2 are *in-phase* and in the second case they are *anti-phase*:

$$z'_{1-j}(x) = (1 - \sigma - \Phi)f'(z_j(x))z'_j(x) + \frac{\sigma}{2r}[f(z_j(x+r)) - f(z_j(x-r))]$$
$$+ \Phi \begin{cases} f'(z_j(x)) \; z'_j(x) & \text{in-phase} \\ f'(z_{1-j}(x)) \; z'_{1-j}(x) & \text{anti-phase.} \end{cases} \tag{7.20}$$

At the point x^* where the smooth profile breaks up and chimera states are born, the spatial derivative becomes infinite. Considering that $z'_0(x^*)$, $z'_1(x^*)$ diverge to infinity, we neglect the term without derivative $\frac{\sigma}{2r}[f(z_j(x+r)) - f(z_j(x-r))]$. Therefore, we obtain for the odd and even time steps:

$$j = 0: \quad z_1'(x^*) = (1 - \sigma - \Phi) f'(z_0(x^*)) z_0'(x^*)$$

$$+ \, \Phi \begin{cases} f'(z_0(x^*)) \, z_0'(x^*) & \text{in-phase} \\ f'(z_1(x^*)) \, z_1'(x^*) & \text{anti-phase,} \end{cases} \tag{7.21}$$

$$j = 1: \quad z_0'(x^*) = (1 - \sigma - \Phi) f'(z_1(x^*)) z_1'(x^*)$$

$$+ \, \Phi \begin{cases} f'(z_1(x^*)) \, z_1'(x^*) & \text{in-phase} \\ f'(z_0(x^*)) \, z_0'(x^*) & \text{anti-phase.} \end{cases} \tag{7.22}$$

Since we choose the parameters for every layer in the regime of wave number $k = 1$ and period-2 dynamics in time, we assume that the spatial derivatives at even and odd time steps satisfy $z_0'(x) = -z_1'(x)$ and that at the break-up point $z_0(x^*) = z_1(x^*) \equiv z^*$ and $z_0'(x^*) \equiv z'^*$ holds. Multiplying the upper equations (in-phase) for $j = 0$ and $j = 1$ we obtain for the two cases:
(i) First and second layer in-phase:

$$(z'^*)^2 = (1 - \sigma - \Phi)^2 f'(z^*)^2 \, (z'^*)^2 + 2\Phi \, (1 - \sigma - \Phi) f'(z^*)^2 \, (z'^*)^2$$

$$+ \, \Phi^2 f'(z^*)^2 \, (z'^*)^2 \tag{7.23}$$

which yields

$$1 = (1 - \sigma)^2 f'(z^*)^2. \tag{7.24}$$

(ii) First and second layer anti-phase:

$$(z'^*)^2 = (1 - \sigma - \Phi)^2 f'(z^*)^2 \, (z'^*)^2 - 2\Phi \, (1 - \sigma - \Phi) f'(z^*)^2 \, (z'^*)^2$$

$$+ \, \Phi^2 f'(z^*)^2 \, (z'^*)^2 \tag{7.25}$$

which yields

$$1 = (1 - \sigma - 2\Phi)^2 f'(z^*)^2. \tag{7.26}$$

To derive an approximation for the critical coupling strength where the smooth profile breaks up, let z^* be the fixed point of the local logistic map: $z^* = f(z^*) = az^*(1 - z^*)$ hence $z^* = 1 - 1/a \approx 0.737$ with $a = 3.8$. We solve Eqs. (7.24) and (7.26) with $f'(z^*) = a(1 - 2z^*) = 2 - a$:
(i) in-phase:

$$1 - \sigma = \pm \frac{1}{|f'(z^*)|}, \tag{7.27}$$

ii) anti-phase:

$$1 - \sigma - 2\Phi = \pm \frac{1}{|f'(z^*)|}. \tag{7.28}$$

We choose the plus sign of Eqs. (7.27) and (7.28) since the lower value of σ represents the threshold where the smooth profile breaks up with decreasing coupling strength. With $|f'(z^*)| = 1.8$, we derive the condition for the onset of phase chimeras if layers 1 and 2 are in:

$$\left. \begin{matrix} \text{in-phase:} & \sigma \\ \text{anti-phase:} & \sigma + 2\Phi \end{matrix} \right\} \approx 0.44. \tag{7.29}$$

Thus phase chimeras only exist if the effective coupling strength is below this critical value, i.e., $\sigma < 0.44$ if layers 1 and 2 oscillate in-phase, and $\sigma + 2\Phi < 0.44$ if the two layers oscillate anti-phase. Interestingly, for in-phase oscillation the inter-layer coupling Φ has no effect on the critical coupling strength. On the other hand, if we consider anti-phase oscillations, the inter-layer coupling strength Φ has to be taken into account. This explains why phase chimeras can exist in single layers, but with increasing inter-layer coupling strength they disappear. Of course the condition $\sigma + 2\Phi < 0.44$ has been derived under very crude approximations and therefore only a rough estimate like $\sigma + 2\sigma_{ij} < 0.4$ can be applied. Indeed, in Figs. 7.19 and 7.20, we find phase chimeras only for small σ_{ij}, as in the marked point A of Fig. 7.19 where $\sigma_i + 2\sigma_{ij} = 0.25$. In point C, i.e., the right column of Fig. 7.20 with $\sigma_i + 2\sigma_{ij} = 0.49$ we cannot observe phase chimeras anymore, and point B, i.e., the middle column with $\sigma_i + 2\sigma_{ij} = 0.35$ seems to correspond to a transition zone of both regimes within the possible error of our approximation.

7.4.3   Comparison with Time-Continuous Systems

We have demonstrated that multiplex networks of time-discrete maps allow for intriguing relay synchronization scenarios, where distant layers synchronize in spite of the absence of direct connections between them. We have analyzed relay synchronization in a three-layer network of logistic maps, with nonlocal coupling topologies within the layers. The uncoupled nodes are characterized by chaotic dynamics, however, due to the coupling the logistic map gives rise to spatially smooth profiles or chimera patterns, and can perform periodic dynamics in time. In contrast to the time-continuous systems in Sects. 7.1, 7.2 and 7.3, we observe two types of chimera states with large (phase chimeras) and small (amplitude chimeras) variations of the dynamical variable, and demonstrate the transition from one type of chimera (phase chimeras) to the other with increasing inter-layer coupling strength. A similar transition depending on the coupling strength has been analyzed in Sect. 4.3.4: There, the occurrence of amplitude chimeras in a single-layer ring network depends linearly on the coupling strength. In multiplex networks of time-continuous oscillators, we have detected the scenario of synchronized phase chimeras. In comparison with time-discrete maps, we find regimes of relay synchronization between amplitude chimeras in the outer layers, as well as partial relay synchronization of chimera states in the two outer layers in the form of intriguing double chimeras, where the coherent domains in both layers are synchronized, while the incoherent ones are not. By choosing an appropriate value for the inter-layer coupling strength we can switch between the different synchronization scenarios.

On balance, we confirmed the phenomenon of (partial) relay synchronization to be independent of local dynamics in the multiplex network. Moreover, we have provided an analytic approximation for coupling strengths of the network necessary for phase chimeras, and have explained with this our observation that phase chimeras disappear with increasing inter-layer coupling strength. Our findings show that weak inter-layer coupling with small σ_{ij} is crucial for relay and partial relay synchronization in the networks. The advantage of this simple paradigmatic model over complex oscillator models is, that it allows for analytical insight into the dynamics of patterns.

7.5 Mechanism Behind Partial Relay Synchronization

Synchronization of chimera states is a current research topic [5, 6]. Moreover, partial relay synchronization of chimera states is a phenomenon where, on the one hand complex dynamics arise on complex networks, on the other hand these dynamics start to behave partially synchronized. It is difficult to develop analytical instruments to explain this complex behavior. Nevertheless, in the present Section, we want to give an explanatory approach by taking a closer look on the coupling terms.

7.5.1 Relay Synchronization in 3-Node Networks

There is some analogy between the full relay synchronization in a 3-layer network and in a 3-node network of delay-coupled lasers [71], see also [26], where the similarity between an active and a passive relay is elaborated, and intuitively explained by two delayed feedback loops which describe the passive relay (semitransparent mirror). Taking into account the so-called quotient network behind the three-layer network one intuitively expects the outer layers to behave the same way because of the underlying symmetry. Quotient networks have been introduced to transform certain large networks into a reduced system, which can be then treated analytically [1]. Extending the 3-node network to the much more complex phenomenon of partial relay synchronization of chimera states has been a motivation for the present study. A simple three-node relay network cannot reproduce this phenomenon. In Fig. 7.5b a classical scenario of relay synchronization is shown. The coherent parts of the chimera states in the three layers are synchronized in such a way, that the outer lay ers are in-phase synchronized, whereas the relay layer is in anti-phase synchronized. For most of the nodes in the coherent domain the intra-layer coupling term vanishes due to the finite coupling range R_i and the diffusive coupling. Therefore, the synchronization mechanism can be compared to a system of three coupled oscillators in [26].

7.5.2 Intrinsic Pacemaker

In Sect. 7.2, we have analyzed the relay synchronization tongues in Fig. 7.4. Starting form partial relay synchronization (red region) it is possible to achieve full relay synchronization by increasing the inter-layer coupling σ_{ij}. In Fig. 7.22, the detailed transition from partial relay (double chimera) to full relay synchronization is shown. By increasing the inter-layer coupling the arc-shaped profile of the local inter-layer synchronization error E_k^{13} for $\sigma_{ij} = 0.013$ gets a growing bump in the center of the incoherent part of the chimera state. This means that synchronization starts from the middle of the incoherent domain and then spreads to its boarders ($\sigma_{ij} = 0.025$). In Fig. 7.22 for $\sigma_{ij} = 0.027$, the chimera states in the outer layers are fully synchronized except the region between coherent and incoherent domains. For $\sigma_{ij} = 0.029$ in Fig. 7.22, the two bumps vanish and full relay synchronization can be achieved.

In Chap. 4, we have pointed out the importance of initial conditions for chimera states by analyzing the coupling function of the single oscillators. The provided arguments hold also in case of chimera states in multiplex networks. For a coupling radius $R_i < N/2$ and in case of a chimera state in layer i, the dynamics of the nodes in the middle of the coherent domain are specific. Because of the (undelayed) diffusive scheme of the intra-layer coupling term in Eq. (7.6), the contribution of this term for the dynamics vanishes. The nodes have no input from the intra-layer coupling term. For large systems $N \to \infty$ and a corresponding coupling radius R_i, this argumentation holds also for the middle of the incoherent region of the chimera state. The contributions of the incoherent neighbors cancel each other and also the nodes in the middle of the incoherent region have no input from the intra-layer coupling term. In Fig. 7.23 by way of example, the intra-layer coupling term is depicted for chimera states in two layers of a network: In the middle of the coherent and incoherent domains of the chimera states the absolute value of the intra-layer coupling term has its minimum (light blue color).

The authors of [54] have recently proposed the concept of a pacemaker oscillator in a ring network. They have modified the coupling topology by cutting all incoming links of one oscillator, the so-called pacemaker. By doing so, they have not only stabilized chimera states, but also prevented their drifting motion, e.g., as shown in Fig. 7.12c. Interestingly, the pacemaker attracts the incoherent domain of chimera states and is located in its center. For synchronization of chimera states in multiplex networks the center of the incoherent and coherent domain plays an important role. In general, with the increasing of the inter-layer coupling the coherent domains of the chimera states in the layers start to synchronize. A synchronized chimera state

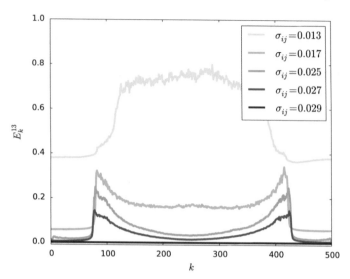

Fig. 7.22 Transition from partial relay (double chimera) to full relay synchronization: local inter layer synchronization error between the outer layers E_k^{13} in dependence on the node index k for fiv different values of $\sigma_{ij} = [0.013, 0.017, 0.025, 0.027, 0.029]$. The relay layer is behaving simila to Fig. 7.5b ($E_k^{12} > 2$). For all simulations of the full Eq. (7.6) random initial conditions are used Other parameters are chosen as $\tau = 0.5$, $\sigma_i = 0.2$, $\varepsilon = 0.05$, $a = 0.5$, $N = 500$, $R_i = 170$ $\phi = \frac{\pi}{2} - 0.1$, and $i, j = 1, 2, 3$

between two layers is exemplarily shown in Fig. 7.23b or by way of numerical sim ulations in Fig. 7.13b. Nevertheless, the scenario of "anti-synchronized" dynamic as shown in Fig. 7.23a is possible, where the coherent domain in one layer starts t synchronize with the incoherent domain in the other layer. Numerical simulation confirm this scenario as shown in Fig. 7.13a (between two neighboring layers). Th centers of incoherent and coherent domain act in a sort of an intrinsic pacemake Therefore, the synchronization mechanism can be understood as the interaction c the intrinsic pacemakers between the layers, comparable to the relay synchronizatio of three nodes [26, 27, 59, 60]. In [6], synchronization between chimera states ha been investigated in a network with two layers. In contrast to the multiplex struc ture in Fig. 7.23, the authors have implemented a mean field coupling between tw non-local rings of different size and analyzed synchronization scenarios.

7.6   Summary

In this Chapter, we have shown that multilayer networks allow for intriguing remot synchronization scenarios. Relay synchronization of chimeras between the oute layers of a multiplex network is an example of such a nontrivial scenario, whe

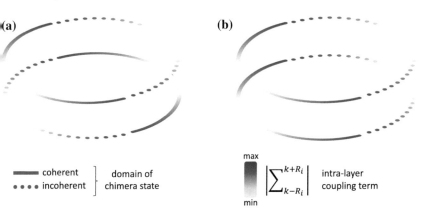

Fig. 7.23 Two synchronization scenarios of chimera states: Each panel shows exemplarily the synamics of a multiplex network with two layers (circles). The dynamics in each layer are given by chimera state, whereby the solid lines denote the coherent domains of the chimera states, whereas e dotted lines stand for the incoherent domains. The brightness of the blue color of the lines indicates the absolute value of the intra-layer coupling term. Panel **a** shows "anti-synchronized" chimera states, where the incoherent domain of one layer is located at the position of the coherent omain of the other layer, and vice versa. Panel **b** depicts synchronized chimera states, where the coherent domain of one layer is located at the position of the incoherent domain of the other layer, and vice versa. Index k denotes the node index and R_i the non-local coupling range in layer i

istant layers of the network synchronize in spite of the absence of direct connections etween them. We have analyzed relay synchronization in a three-layer network of itzHugh-Nagumo oscillators, with nonlocal coupling topologies within the layers, ad have extended the notion of relay synchronization to chimera states.

Chimera patterns, consisting of coherent and incoherent domains, can be observed each network layer; they are usually strongly dependent on the initial conditions, ad it is not possible to predict which part of the network will form coherent domains. y relay synchronization we can fix the location to the same position as in the other iter layer. Varying the strength of the coupling between the network layers, we bserve various scenarios of synchronization of chimera states, either in all three yers, or only in the two outer layers. As measures we employ the global and local iter-layer synchronization errors and mean phase velocity profiles of the oscillators. urthermore, to investigate the dependence on the local dynamics we have chosen ring of logistic maps and successfully validated the phenomenon of partial relay nchronization.

Time delay in the inter-layer coupling, which is ubiquitous in real-world systems ad reflects transmission delays between layers, has been identified as a powerful tool or control of the patterns: It allows for observation of novel synchronization scenar-s where the coherent domains of chimera states in the outer layers are synchronized, hile the incoherent domains are not. The relay layer remains desynchronized and chibits various multi-chimera patterns, or even chaotic dynamics. Furthermore, par-al relay synchronization of chimeras states in the two outer layers has been realized in the form of intriguing double chimeras, where the coherent domains in both layer are synchronized, while the incoherent ones are not. By choosing an appropriate value for the time delay we can switch between the different synchronization scenarios. Moreover, in the the interplay of inter- and intra-layer time delay full and relay inter-layer synchronization can be found and controlled. Control of chimera pattern can also be affected by changing the topology in the intermediate layer. By varying the coupling range we find that even strongly diluted relay layers allow for remote synchronization of chimeras in the outer layers, while the relay layer stays in the chaotic regime.

We propose that our findings may be useful in the study of novel concepts for encrypted and secure communication, where relay synchronization of complex spatio-temporal patterns, for instance chimera states, can be employed. Since the dynamics of the intermediate (relay) layer is not synchronized, it does not transmit information to someone listening in. While relay synchronization of single chaotic laser has been extensively investigated in the context of encrypted communication (see the review paper [71]), here we have extended and generalized the concept of relay synchronization to multilayer networks, which exhibit much more complex dynamics. As brain networks are often described as multilayer structures, our results may also help in elucidating complex scenarios of information processing in neural networks. Recent research in neuroscience indicates that many parts of the brain, e.g. thalamus, interneurons, and hippocampus, act as a relay that connects two different regions [33, 37, 73, 74]. In particular, we have shown, going far beyond the usual complete relay synchronization in multilayer networks, that specifically the hitherto unexplained experiments on imperfect synchronization in the mice brain [32] might be explained by our novel scenarios of partial relay inter-layer synchronization (see Sect. 7.3.4). Our results can be useful for the analysis of relay synchronization in multiplex and multilayer networks with more complex dynamics of the individual nodes, and thus for numerous applications where relay synchronization occurs, e.g. in the mammalian or human brain.

References

1. Aguiar MAD, Dias APS, Golubitsky M, Leite MCA (2009) Bifurcations from regular quotient networks: a first insight. Phys D **238**:137–155
2. Ahlborn A, Parlitz U (2007) Controlling spatiotemporal chaos using multiple delays. Phys Rev E **75**:65202
3. Albert R, Barabási AL (2002) Statistical mechanics of complex networks. Rev Mod Phys **74**:47–97
4. Andrzejak RG, Rummel C, Mormann F, Schindler K (2016) All together now: analogies between chimera state collapses and epileptic seizures. Sci Rep **6**:23000
5. Andrzejak RG, Ruzzene G, Malvestio I (2017) Generalized synchronization between chimera states. Chaos **17**:053114
6. Andrzejak RG, Ruzzene G, Malvestio I, Schindler K, Schöll E, Zakharova A (2018) Mean field phase synchronization between chimera states. Chaos **28**:091101

7. Ashwin P, Timme M (2005) Unstable attractors: existence and robustness in networks of oscillators with delayed pulse coupling. Nonlinearity **18**:2035
8. Atay FM (ed) (2010) Complex time-delay systems, understanding complex systems. Springer, Berlin, Heidelberg
9. Bachmair CA, Schöll E (2014) Nonlocal control of pulse propagation in excitable media. Eur Phys J B **87**:276
10. Banerjee T, Ghosh D, Biswas D, Schöll E, Zakharova A (2018) Networks of coupled oscillators: from phase to amplitude chimeras. Chaos **28**:113124
11. Banerjee T, Bandyopadhyay B, Zakharova A, Schöll E (2019) Filtering suppresses amplitude chimeras. Front Appl Math Stat **5**:8
12. Boccaletti S, Bianconi G, Criado R, del Genio CI, Gómez-Gardeñes J, Romance M, Sendiña Nadal I, Wang Z, Zanin M (2014) The structure and dynamics of multilayer networks. Phys Rep **544**:1–122
13. Bogomolov S, Strelkova G, Schöll E, Anishchenko VS (2016) Amplitude and phase chimeras in an ensemble of chaotic oscillators. Tech Phys Lett **42**:765–768
14. Bogomolov S, Slepnev A, Strelkova G, Schöll E, Anishchenko VS (2017) Mechanisms of appearance of amplitude and phase chimera states in a ring of nonlocally coupled chaotic systems. Commun Nonlinear Sci Numer Simul **43**:25
15. Bordyugov G, Pikovsky A, Rosenblum M (2010) Self-emerging and turbulent chimeras in oscillator chains. Phys Rev E **82**:035205
16. Bountis T, Kanas V, Hizanidis J, Bezerianos A (2014) Chimera states in a two-population network of coupled pendulum-like elements. Eur Phys J Spec Top **223**:721–728
17. Brandstetter SA, Dahlem MA, Schöll E (2010) Interplay of time-delayed feedback control and temporally correlated noise in excitable systems. Philos Trans Royal Soc A **368**:391
18. Bukh A, Rybalova E, Semenova N, Strelkova G, Anishchenko V (2017) New type of chimera and mutual synchronization of spatiotemporal structures in two coupled ensembles of nonlocally interacting chaotic maps. Chaos **27**:111102
19. Bukh AV, Slepnev A, Anishchenko VS, Vadivasova TE (2018) Stability and noise-induced transitions in an ensemble of nonlocally coupled chaotic maps. Reg Chaotic Dyn **23**:325–338
20. Buldú JM, Porter MA (2018) Frequency-based brain networks: from a multiplex framework to a full multilayer description. Netw Neurosci **2**:418–441
21. Cholvin T, Hok V, Giorgi L, Chaillan FA, Poucet B (2018) Ventral midline thalamus is necessary for hippocampal place field stability and cell firing modulation. J Neurosci **38**:158
22. Chouzouris T, Omelchenko I, Zakharova A, Hlinka J, Jiruska P, Schöll E (2018) Chimera states in brain networks: empirical neural versus modular fractal connectivity. Chaos **28**:045112
23. De Domenico M (2017) Multilayer modeling and analysis of human brain networks. Gigascience **6**:1–8
24. Feng YE, Li HH (2015) The dependence of chimera states on initial conditions. Chin Phys Lett **32**:060502
25. Fischer I, Vicente R, Buldú JM, Peil M, Mirasso CR, Torrent MC, García-Ojalvo J (2006) Zero-lag long-range synchronization via dynamical relaying. Phys Rev Lett **97**:123902
26. Flunkert V, D'Huys O, Danckaert J, Fischer I, Schöll E (2009) Bubbling in delay-coupled lasers. Phys Rev E **79**:065201(R)
27. Flunkert V (2011) Delay-coupled complex systems, Springer Theses. Springer, Heidelberg
28. Geffert PM, Zakharova A, Vüllings A, Just W, Schöll E (2014) Modulating coherence resonance in non-excitable systems by time-delayed feedback. Eur Phys J B **87**:291
29. Ghosh S, Kumar A, Zakharova A, Jalan S (2016) Birth and death of chimera: interplay of delay and multiplexing. Europhys Lett **115**:60005
30. Ghosh S, Jalan S (2018) Engineering chimera patterns in networks using heterogeneous delays. Chaos **28**:071103
31. Gjurchinovski A, Schöll E, Zakharova A (2017) Control of amplitude chimeras by time delay in dynamical networks. Phys Rev E **95**:042218
32. Gollo LL, Mirasso CR, Atienza M, Crespo-Garcia M, Cantero JL (2011) Theta band zero-lag long-range cortical synchronization via hippocampal dynamical relaying. PLoS ONE **6**:e17756

33. Guillery RW, Sherman SM (2002) Thalamic relay functions and their role in corticocortical communication: generalizations from the visual system. Neuron **33**:163–175
34. Hövel P, Schöll E (2005) Control of unstable steady states by time-delayed feedback methods. Phys Rev E **72**:046203
35. Hagerstrom AM, Murphy TE, Roy R, Hövel P, Omelchenko I, Schöll E (2012) Experimental observation of chimeras in coupled-map lattices. Nat Phys **8**:658–661
36. Halverson E, Poremba A, Freeman JH (2015) Medial auditory thalamus is necessary for acquisition and retention of eyeblink conditioning to cochlear nucleus stimulation. Learn Mem **22**:258
37. Halassa MM, Kastner S (2017) Thalamic functions in distributed cognitive control. Nat Neurosci **20**:1669–1679
38. Havlin S, Kenett DY, Ben-Jacob E, Bunde A, Cohen R, Hermann H, Kantelhardt JW, Kertész J, Kirkpatrick S, Kurths J, Portugali J, Solomon S (2012) Challenges in network science: applications to infrastructures, climate, social systems and economics. Eur Phys J Spec Top **214**:273–293
39. Jalan S, Sarkar C, Madhusudanan A, Dwivedi SK (2014) Uncovering randomness and success in society. PloS one **9**:1–8
40. Kemeth FP, Haugland SW, Schmidt L, Kevrekidis YG, Krischer K (2016) A classification scheme for chimera states. Chaos **26**:094815
41. Kivelä M, Arenas A, Barthélemy M, Gleeson JP, Moreno Y, Porter MA (2014) Multilayer networks. J Complex Netw **2**:203–271
42. Kuramoto Y, Battogtokh D (2002) Coexistence of coherence and incoherence in nonlocally coupled phase oscillators. Nonlin Phen Complex Sys **5**:380–385
43. Leyva I, Sendiña-Nadal I, Sevilla-Escoboza R, Vera-Avila VP, Chholak P, Boccaletti S (2018) Relay synchronization in multiplex networks. Sci Rep **8**:8629
44. Majhi S, Perc M, Ghosh D (2017) Chimera states in a multilayer network of coupled and uncoupled neurons. Chaos **27**:073109
45. Martens EA, Laing CR, Strogatz SH (2010) Solvable model of spiral wave chimeras. Phys Rev Lett **104**:044101
46. Masoliver M, Malik N, Schöll E, Zakharova A (2017) Coherence resonance in a network of FitzHugh-Nagumo systems: interplay of noise, time-delay and topology. Chaos **27**:101102
47. Motter AE (2010) Nonlinear dynamics: spontaneous synchrony breaking. Nat Phys **6**:164–165
48. Omel'chenko OE, Wolfrum M, Maistrenko Y (2010) Chimera states as chaotic spatiotemporal patterns. Phys Rev E **81**:065201(R)
49. Omelchenko I, Maistrenko Y, Hövel P, Schöll E (2011) Loss of coherence in dynamical networks: spatial chaos and chimera states. Phys Rev Lett **106**:234102
50. Omelchenko I, Riemenschneider B, Hövel P, Maistrenko Y, Schöll E (2012) Transition from spatial coherence to incoherence in coupled chaotic systems. Phys Rev E **85**:026212
51. Omel'chenko OE, Wolfrum M, Yanchuk S, Maistrenko Y, Sudakov O (2012) Stationary patterns of coherence and incoherence in two-dimensional arrays of non-locally-coupled phase oscillators. Phys Rev E **85**:036210
52. Panaggio MJ, Abrams DM (2015) Chimera states: coexistence of coherence and incoherence in networks of coupled oscillators. Nonlinearity **28**:R67
53. Rothkegel A, Lehnertz K (2014) Irregular macroscopic dynamics due to chimera states in small-world networks of pulse-coupled oscillators. New J Phys **16**:055006
54. Ruzzene G, Omelchenko I, Schöll E, Zakharova A, Andrzejak RG (2019) Controlling chimera states via minimal coupling modification. Chaos (preprint)
55. Sarkar C, Yadav A, Jalan S (2016) Multilayer network decoding versatility and trust. Europhys Lett **113**:18007
56. Sawicki J, Omelchenko I, Zakharova A, Schöll E (2018) Synchronization scenarios of chimeras in multiplex networks. Eur Phys J Spec Top **227**:1161
57. Sawicki J, Omelchenko I, Zakharova A, Schöll E (2018) Delay controls chimera relay synchronization in multiplex networks. Phys Rev E **98**:062224

58. Sawicki J, Ghosh S, Jalan S, Zakharova A (2019) Chimeras in multiplex networks: interplay of inter- and intra-layer delays. Front Appl Math Stat **5**:19
59. Schöll E, Hiller G, Hövel P, Dahlem MA (2009) Time-delayed feedback in neurosystems. Phil Trans R Soc A **367**:1079–1096
60. Schöll E, Hövel P, Flunkert V, Dahlem MA (2010) Time-delayed feedback control: from simple models to lasers and neural systems. In: Atay FM (ed) Complex time-delay systems: theory and applications. Springer, Berlin, pp 85–150
61. Semenov V, Feoktistov A, Vadivasova T, Schöll E, Zakharova A (2015) Time-delayed feedback control of coherence resonance near subcritical Hopf bifurcation: theory versus experiment. Chaos **25**:033111
62. Semenova N, Zakharova A, Schöll E, Anishchenko VS (2015) Does hyperbolicity impede emergence of chimera states in networks of nonlocally coupled chaotic oscillators? Europhys Lett **112**:40002
63. Semenov V, Zakharova A, Maistrenko Y, Schöll E (2016) Delayed-feedback chimera states: forced multiclusters and stochastic resonance. Europhys Lett **115**:10005
64. Semenova N, Zakharova A, Anishchenko VS, Schöll E (2016) Coherence-resonance chimeras in a network of excitable elements. Phys Rev Lett **117**:014102
65. Semenova N, Strelkova G, Anishchenko VS, Zakharova A (2017) Temporal intermittency and the lifetime of chimera states in ensembles of nonlocally coupled chaotic oscillators. Chaos **27**:061102
66. Sethia GC, Sen A (2014) Chimera states: the existence criteria revisited. Phys Rev Lett **112**:144101
67. Shepelev IA, Bukh AA, Vadivasova TE, Anishchenko VS, Zakharova A (2017) Double-well chimeras in 2D lattice of chaotic bistable elements. Commun Nonlinear Sci Numer Simul **54**:50–61
68. Shepelev IA, Bukh AV, Strelkova GI, Vadivasova TE, Anishchenko VS (2017) Chimera states in ensembles of bistable elements with regular and chaotic dynamics. Nonlinear Dyn **90**:2317
69. Singh A, Jalan S, Kurths J (2013) Role of delay in the mechanism of cluster formation. Phys Rev E **87**:030902(R)
70. Singh A, Ghosh S, Jalan S, Kurths J (2015) Synchronization in delayed multiplex networks. Europhys Lett **111**:30010
71. Soriano MC, García-Ojalvo J, Mirasso CR, Fischer I (2013) Complex photonics: dynamics and applications of delay-coupled semiconductors lasers. Rev Mod Phys **85**:421–470
72. Vadivasova TE, Strelkova G, Bogomolov SA, Anishchenko VS (2016) Correlation analysis of the coherence-incoherence transition in a ring of nonlocally coupled logistic maps. Chaos **26**:093108
73. Vann SD, Nelson AJD (2015) The mammillary bodies and memory: more than a hippocampal relay. In: Progress in brain research, vol 219. Elsevier, Amsterdam, pp 163–185
74. Wang X, Vaingankar V, Sanchez CS, Sommer FT, Hirsch JA (2011) Thalamic interneurons and relay cells use complementary synaptic mechanisms for visual processing. Nat Neurosci **14**:224
75. Winkler M (2018) Synchronization of chimera states in multiplex networks of logistic maps. Bachelor's thesis, Technische Universität Berlin
76. Winkler M, Sawicki J, Omelchenko I, Zakharova A, Anishchenko V, Schöll E (2019) Relay synchronization in multiplex networks of discrete maps. Europhys Lett **126**:50004
77. Wolfrum M, Omel'chenko OE, Yanchuk S, Maistrenko Y (2011) Spectral properties of chimera states. Chaos **21**:013112
78. Yanchuk S, Wolfrum M, Hövel P, Schöll E (2006) Control of unstable steady states by long delay feedback. Phys Rev E **74**:026201
79. Yanchuk S, Perlikowski P (2009) Delay and periodicity. Phys Rev E **79**:046221
80. Zakharova A, Kapeller M, Schöll E (2014) Chimera death: symmetry breaking in dynamical networks. Phys Rev Lett **112**:154101
81. Zakharova A, Semenova N, Anishchenko VS, Schöll E (2017) Time-delayed feedback control of coherence resonance chimeras. Chaos **27**:114320

Chapter 8
Conclusion

8.1 Summary

The focus of this thesis has been the study of synchronization phenomena in complex networks and their control through time delay. Starting from a pair of oscillators and proceeding via simple ring networks, we have outlined the progression to complex multilayer structures, examining synchronization in many of its facets. Thereby, Part I has been focused on the study of synchronization phenomena in single-layer systems and Part II has explored synchronization scenarios in multilayer networks as a common description of neuronal brain structures.

After the general concepts have been discussed and the necessary background to complex networks has been given in Chap. 2, we have started to investigate the synchronization of two coupled oscillators using the example of organ pipes (Part I). In Chap. 3, we have used a nonlinear oscillator as a simplified model for organ pipes, i.e., two Van der Pol oscillators with delayed coupling. To gain a deeper understanding of the various bifurcation scenarios, we have developed and extended two complementary analytical approaches: The method of averaging, on the one hand, has allowed us to obtain a generalized Adler equation for the phase dynamics, which could then be used to study the stability of the equilibria corresponding to frequency locking of the oscillators. On the other hand, the describing function method has allowed us to determine the synchronization frequency and, hence, explain the curvature of the frequency versus detuning. The latter is found in both, the numerical simulations, as well as in the experiments. A detailed bifurcation analysis has affirmed the existence of in- and anti-phase synchronization. In each case the synchronization frequency has a different value which is in perfect accordance with our analytic calculations. By introducing a delay-dependent coupling strength $\kappa(\tau)$, we can moreover explain the details of the experimentally observed Arnold tongue.

In Chap. 4, we have analyzed special partial synchronization patterns called chimera states in ring networks with nonlocal coupling. These patterns have been intensively investigated over the last fourteen years but analytic insights about their emergence were still missing. We have provided an analytical argument to explain

© Springer Nature Switzerland AG 2019

Sawicki, *Delay Controlled Partial Synchronization in Complex Networks*,

Springer Theses, https://doi.org/10.1007/978-3-030-34076-6_8

the need for an off-diagonal coupling, i.e., a phase-lag in the coupling, in order to create chimera states. Based on this, we have discussed the impact of the sign of the coupling phases. We could further show how the sign of the coupling phase determines the sign of the profile of the mean phase velocities. Furthermore, we exemplified how our argument gives an intuitive explanation for the transition from phase chimera states, in the limit of weak coupling to a state sharing the main features of an amplitude-mediated chimera state, in the case of intermediate coupling strength.

In Chap. 5, we have analyzed chimera states in ring networks of oscillators with hierarchical connectivities, as arising in neuroscience, and provided a numerical and analytical study of complex spatio-temporal patterns. In comparison with nonlocal topologies, fractal ones allow for less links to obtain chimera states. Moreover, our study has been focused on the role of time delay in the coupling term and its ability to stabilize several types of chimera states. We have demonstrated that time delay can induce patterns which are not observed in the undelayed case. To prove the generality of our observations we have made use of several paradigmatic models for the oscillators. In addition, we have analytically investigated the influence of the delay time τ upon the period; i.e., the phase velocity. We have found a piecewise linear dependence in regimes with coherent states, whereas a nonlinear dependence upon τ is found for incoherent states.

Part II has been devoted to synchronization phenomena in multilayer network which can sufficiently describe many real world scenarios. In Chap. 6, we have considered empirical connectivity matrices measured in the human brain. The brain has naturally a hemispheric structure, which can be modeled in the multilayer framework two hemispheres with a coupling within and between them. A general, structural asymmetry in the brain allows for partial synchronization dynamics, which can be used to model unihemispheric sleep or explain the mechanism for first-night effects. By analyzing partial synchronization patterns we have specified the conditions for dynamical asymmetries of the hemispheres and reported unihemispheric sleep-like patterns regarding the transition from frequency to in-phase synchronization. Such intriguing dynamical patterns and their synchronization or desynchronization across layers of a multilayer network can have high impact on technology. Moreover, studying these phenomena may allow us to better understand highly complex systems such as real neural populations in the mammalian brain.

In Chap. 7, we have demonstrated that multilayer networks allow for intriguing remote synchronization scenarios by extending the notion of relay synchronization from single oscillators to chimera states, which we have analyzed in an ordinary triplex network. By relay synchronization we can fix the location of the coherent domains to the same position as in the other outer layer. Varying the strength of the coupling between the network layers, we observe various scenarios of chimera synchronization, either in all three layers, or only in the two outer layers. To quantify the states, we have introduced the global and local inter-layer synchronization error. Furthermore, to investigate the dependence on the local dynamics, we have chosen a multiplex network of logistic maps and successfully validated the phenomenon of partial relay synchronization. Time delay in the inter-layer coupling has been identified as a powerful tool for control of the patterns: It allows for observation of novel synchronization scenarios in the form of intriguing double chimeras where the coherent domains of chimera states in the outer layers are synchronized, while the incoherent domains are not. The relay layer remains desynchronized and exhibits various multi-chimera patterns, or even chaotic dynamics.

Our findings may be useful in the study of novel concepts for encrypted and secure communication, where relay synchronization of complex spatio-temporal patterns, for instance chimera states, can be employed. Since the dynamics of the intermediate (relay) layer is not synchronized, it does not transmit information to someone listening in. As brain networks are often described as multilayer structures, our results may further help in elucidating complex scenarios of information processing in neural networks. Recent research in neuroscience indicates that many parts of the brain, e.g., thalamus, interneurons, and hippocampus, act as a relay that connects two different regions [10, 11, 16, 18]. In particular, we have shown, going far beyond the usual complete relay synchronization in multilayer networks, that specifically the hitherto unexplained experiments on imperfect synchronization in the mice brain [8] might be explained by our novel scenarios of partial relay inter-layer synchronization (see Sect. 7.3.4).

In summary, we have investigated partial synchronization patterns in complex networks. We have used the insights gained by first studying a simple two-node system to explain the emerging complexity in case of larger networks. We have introduced multilayer networks and observed a novel synchronization scenarios in the form of double chimeras. We have identified delay time in the coupling terms as a powerful tool to control partial synchronization. Furthermore, we have analyzed these synchronization patterns in the interplay of topology and various dynamical models, which are of particular interest for applications in neurosciences.

8.2 Outlook

At the interface of musicology and physics we have analyzed the synchronization and bifurcation scenarios of two organ pipes. Chapter 3 has contributed substantially explaining these phenomena and provided a fundamental understanding. In particular, time delay has been found to play a crucial role for stability of dynamical states. For the sake of simplicity, the organ pipe has been modeled by a single Van der Pol oscillator with an intrinsic frequency. In general, musical instruments are characterized by their harmonic spectra. Therefore, one focus of further research should be to implement multiples of the fundamental frequency in the model and to analyze the synchronization scenarios of the single harmonics. Especially the "cross"-synchronization of the fundamental tone with its harmonics could be of great interest; not only for organ voices as "mixtures", where the overtone spectrum is implemented technically. A second focus of further research should include more than two oscillators: One could consider the interplay of a group of pipes, on the one hand, and self-delayed loops caused by reflection, on the other hand.

In Chap. 4, we have elaborated the importance of the interplay of initial conditions and coupling terms for chimera states. Nevertheless, the potential of the provided analytics has not been exploited completely. One possibility is to extend the analytical arguments to more complex topologies as in Chap. 5. Furthermore, the influence of time delay can be taken into account: Due to the equal phase velocity in the coherent domain of the chimera state, one could reduce the dynamics in that domain to an effective problem of one oscillator with a self-delayed loop. In principal, I believe that a further investigation of the coupling terms [5, 6] is crucial to solve the puzzle behind chimera states.

Investigations of the dynamics on multilayer networks are a recent topic of research. We have provided a detailed analysis on asymmetries in empirical structural connectivities of human brains. In Chap. 6, we have exploited the conditions for unihemispheric slow-wave sleep. Even though oscillators are sufficient to describe synchronization patterns, the activity of these nodes as well as their interaction is fully continuous. One might need to consider a minimum amount of discreteness into the model to study questions of neuroscience. As in the nervous system discrete signals (action potentials) and continuous signals (membrane potentials, neuromodulation) are handled differently. Therefore, it would be convenient to introduce a threshold in the oscillator dynamics to define a discrete signal. Another possibility would be to change the local dynamics to either the leaky integrate-and-fire or Hodgkin-Huxley model, where bursting behavior can be observed.

In Chap. 7, we have extended the notion of relay synchronization to chimera states. Using the example of an ordinal triplex network we have developed measures to quantify various synchronization patterns. It is still an open question if relay synchronization is robust with respect to the link-density (number of connections of each node) of the relay layer. To test this aspect, one should change the topology of the relay layer stepwise from local to global and study the local and global synchronization errors between the layers. According to our preliminary results we have found that full relay synchronization seems to be robust for changes in the relay layer density. Secondly, it may be imaginable that the symmetry of the coupling (relay) topology might be more important for relay synchronization than the link density. In this context it is reasonable to study the effects of asymmetric coupling schemes, topologies and clustering in the relay layer. In particular, one could replace the non-local topology by a small world, (semi-)random, or fractal network topology.

Another interesting direction of future research could bring the network topology closer to neuroscientific models to study the delicate interplay of convergence, directed relay transmission and delayed feedback, possibly modeling the thalamocortical loops. Highly symmetric rings of nodes are artificial and therefore not sufficient to describe the nervous system above a certain degree. Next to breaking the artificial symmetry of the system, inducing convergent and divergent pathways in the multi-layer network seems a promising idea: These pathways – in which a layer may not have the same number N of nodes as others – are common in neural information processing. The clearest example is the early visual system where convergence occurs as early as passing visual input from photoreceptors to retinal cells (100–10,000 to 1 depending on eyesight). However, in the same pathway divergence occurs when he relatively small lateral nucleus feeds preprocessed information for the visual cortices. It would be interesting to implement this concept by changing the convergence factor (N_{outer}/N_{relay}) from extreme cases as $N_{relay} = 1$ (relay hub) to fully balanced information flow $N_{relay} = N_{outer}$.

Similarly, it would be interesting to define a specific flow of information throughout the layers by means of information theory. As in the aforementioned example of visual information being transmitted from the retina via the thalamus to the cortex, the inter-layer links could be defined unidirectional ($\sigma_{ij} > \sigma_{ji} = 0$) or unbalanced $\sigma_{ij} > \sigma_{ji} \approx 0$). Of course, in the neuronal system we cannot talk about a clear-cut direction of flow as many areas are fed back to the previous step of information processing (such as cortico-thalamic circuits or loops between the layers of the cortical columns). Therefore, a question would be whether a delayed feedback in an outer layer influences synchronization patterns in an unbalanced network. Examples of such feedback loops can be found between the cortex and the thalamus [1, 4, 4, 7, 9, 13–15, 17]. Secondly, further studies of the robustness could be carried out regarding different inter-layer delays $\tau_{12} \neq \tau_{23}$ in a triplex network. Another approach would be to extent the multiplex network by an adaptive coupling structure inside the single layers similar to [2, 3, 12]. In such networks, the network topology inside a layer changes according to the dynamical state of the nodes, while, in turn, the state of the nodes is influenced by the topology itself. Example of adaptive networks can be found in the synaptic plasticity of the brain or the forming of new links in communication and social networks.

The research on double chimera states is still in an early stage. On the one hand, the challenge is to describe these partial synchronization patterns sufficiently well without requiring too complex node dynamics or network topology. On the other hand, considering more realistic neural models would give us the opportunity to strengthen the applicability and generality of our findings. A first step in this direction has been proposed in Chap. 7, where we have explained experimental results from [8]. Nevertheless, the challenge is to find further analytical explanations and applications to real-world problems.

References

1. Barthó P, Slézia A, Mátyás F, Faradzs-Zade L, Ulbert I, Harris KD, Acsaády L (2014) Ongoing network state controls the length of sleep spindles via inhibitory activity. Neuron **82**:1367
2. Berner R, Fialkowski J, Kasatkin DV, Nekorkin V, Schöll E, Yanchuk S (2019) Self-similar hierarchical frequency clusters in adaptive networks of phase oscillators (not published)
3. Berner R, Schöll E, Yanchuk S (2019) Multi-clusters in adaptive networks. SIAM J Appl Dyn Syst
4. Bodor AL, Giber K, Rovó Z, Ulbert I, Acsaády L (2008) Structural correlates of efficient GABAergic transmission in the basal ganglia-thalamus pathway. J Neurosci **28**:3090
5. Bogomolov S, Strelkova G, Schöll E, Anishchenko VS (2016) Amplitude and phase chimeras in an ensemble of chaotic oscillators. Tech Phys Lett **42**:765–768

6. Bogomolov S, Slepnev A, Strelkova G, Schöll E, Anishchenko VS (2017) Mechanisms of appearance of amplitude and phase chimera states in a ring of nonlocally coupled chaotic systems. Commun Nonlinear Sci Numer Simul **43**:25

7. Bokor H, Frere SG, Eyre MD, Slézia A, Ulbert I, Lüthi A, Acsaády L (2005) Selective GABAergic control of higher-order thalamic relays. Neuron **45**:929

8. Gollo LL, Mirasso CR, Atienza M, Crespo-Garcia M, Cantero JL (2011) Theta band zero-lag long-range cortical synchronization via hippocampal dynamical relaying. PLoS ONE **6**:e1775

9. Groh A, Bokor H, Mease RA, Plattner VM, Hangya B, Deschenes M, Acsaády L (2013) Convergence of cortical and sensory driver inputs on single thalamocortical cells. Cereb Cortex **12**:3167

10. Guillery RW, Sherman SM (2002) Thalamic relay functions and their role in corticocortical communication: generalizations from the visual system. Neuron **33**:163–175

11. Halassa MM, Kastner S (2017) Thalamic functions in distributed cognitive control. Nat Neurosci **20**:1669–1679

12. Kasatkin DV, Yanchuk S, Schöll E, Nekorkin VI (2017) Self-organized emergence of multilayer structure and chimera states in dynamical networks with adaptive couplings. Phys Rev E **96**:062211

13. Lavallee P, Urbain N, Dufresne C, Bokor H, Acsaády L, Deschenes M (2005) Feedforward inhibitory control of sensory information in higher-order thalamic nuclei. J Neurosci **25**:748

14. Rovó Z, Ulbert I, Acsaády L (2012) Drivers of the primate thalamus. J Neurosci **32**:17894

15. Rovó Z, Mátyás F, Barthó P, Slézia A, Lecci S, Pellegrini C, Astori S, David C, Hangya B, Lüthi A, Acsaády L (2014) Phasic, nonsynaptic GABA-A receptor-mediated inhibition entrains thalamocortical oscillations. J Neurosci **34**:7137

16. Vann SD, Nelson AJD (2015) The mammillary bodies and memory: more than a hippocampal relay. In: Progress Brain Research, vol 219. Elsevier, Amsterdam, pp 163–185

17. Wanaverbecq N, Bodor AL, Bokor H, Slézia A, Lüthi A, Acsaády L (2008) Contrasting the functional properties of GABAergic axon terminals with single and multiple synapses in the thalamus. J Neurosci **28**:11848

18. Wang X, Vaingankar V, Sanchez CS, Sommer FT, Hirsch JA (2011) Thalamic interneurons and relay cells use complementary synaptic mechanisms for visual processing. Nat Neurosci **14**:224

About the Author

Jakub Sawicki studied music at Listaháskóli Islands University and Berlin University of the Arts. He was granted a Cusanuswerk scholarship and in this context committed himself to social projects in Greenland and India. In course of his studies of physics, he worked on the synchronization of organ pipes at Saint Petersburg State University. Apart from his position at Berlin Cathedral, Jakub Sawicki has deepened his studies of organ improvisation with Prof. Wolfgang Seifen. He has engagements as choir and orchestra conductor and teaches improvisation at Berlin University of the Arts. He also participates in silent movie and improvisation projects. 2010 he was engaged by Theodor-Heuss-Kolleg to compose the music for an international movie production. Jakub Sawicki takes part in master classes and competitions and performs in- and outside Germany.

Apart from the musical activities, Jakub Sawicki has continued his scientific research and received the Dr. rer. nat. degree from the Technical University Berlin in 2019. His work was supervised by Prof. Dr. Dr. h.c. Eckehard Schöll, Ph.D., at the Institute of Theoretical Physics. His research interests include synchronization phenomena, nonlinear delay differential equations, modeling of complex systems and neural dynamics. He has contributed to the following publications:

J. Sawicki, M. Abel, and E. Schöll: *Synchronization in coupled organ pipes*, in *Proceedings of the 7th International Conference on Physics and Control (PhysCon 2015)*, edited by (IPACS Electronic Library, 2015), Istanbul, Turkey.

P. Kalle, J. Sawicki, A. Zakharova, and E. Schöll: *Chimera states and the interplay between initial conditions and non-local coupling*, Chaos **27**, 033110 (2017).

J. Sawicki, I. Omelchenko, A. Zakharova, and E. Schöll: *Chimera states in complex networks: interplay of fractal topology and delay*, Eur. Phys. J. Spec. Top. **226**, 1883–1892 (2017).

Springer Nature Switzerland AG 2019
Sawicki, *Delay Controlled Partial Synchronization in Complex Networks*,
Springer Theses, https://doi.org/10.1007/978-3-030-34076-6

- J. Sawicki, I. Omelchenko, A. Zakharova, and E. Schöll: *Synchronization scenarios of chimeras in multiplex networks*, Eur. Phys. J. Spec. Top. **227**, 1161 (2018).
- J. Sawicki, M. Abel, and E. Schöll: *Synchronization of organ pipes*, Eur. Phys. J B **91**, 24 (2018).
- J. Sawicki, I. Omelchenko, A. Zakharova, and E. Schöll: *Delay controls chimera relay synchronization in multiplex networks*, Phys. Rev. E **98**, 062224 (2018).
- J. Sawicki, I. Omelchenko, A. Zakharova, and E. Schöll: *Delay-induced chimeras in neural networks with fractal topology*, Eur. Phys. J. B **92**, 54 (2019).
- J. Sawicki, S. Ghosh, S. Jalan, and A. Zakharova: *Chimeras in multiplex networks. interplay of inter- and intra-layer delays*, Front. Appl. Math. Stat. **5**, 19 (2019).
- M. Winkler, J. Sawicki, I. Omelchenko, A. Zakharova, V. Anishchenko, and E. Schöll: *Relay synchronization in multiplex networks of discrete maps*, Euro phys. Lett. **126**, 50004 (2019).
- L. Ramlow, J. Sawicki, A. Zakharova, J. Hlinka, J. C. Claussen, and E. Schöll *Partial synchronization in empirical brain networks as a model for unihemispheric sleep*, Europhys. Lett. **126**, 50007 (2019).

Printed in the United States
By Bookmasters